*A Scientist at the Seashore*
*The Moment of Creation*
*The Unexpected Vista*
*Are We Alone?*
(WITH ROBERT T. ROOD)

*From Atoms to Quarks*
*Living in Space*

# Meditations at 10,000 Feet

# Meditations
## at 10,000 Feet

*A Scientist in the Mountains*

## by James Trefil

*Illustrations by Judith Peatross*

CHARLES SCRIBNER'S SONS

NEW YORK

Copyright © 1986 James Trefil

Library of Congress Cataloging-in-Publication Data

Trefil, James S., 1938–
    Meditations at 10,000 feet.

    Includes index.
    1. Geology—Popular works.   2. Mountains—Popular
works.   I. Title.
QE31.T67     1986        551'.0914'3        85-27751
ISBN 0-684-18627-6

Published simultaneously in Canada
by Collier Macmillan Canada, Inc.

Printed in the United States of America


We gratefully acknowledge permission to reprint, on page 228, the
poem by Harold P. Furth, which appeared originally in The New Yorker
on November 10, 1956. Reprinted with permission; © 1956, 1984, The
New Yorker Magazine.

This book is dedicated to my three sons, each of whom has just successfully completed a major task in his life:

James Karel Trefil,
A.B. Princeton '85

Cpl. Stefan James Trefil,
82nd Airborne Division, United States Army

Tomas Jaroslav Waples-Trefil,
age three weeks

# Contents

# Introduction

ONE OF THE WORLD'S best kept secrets is that you don't have to be in a laboratory or a lecture room to learn about science. In a very real sense, the entire world is a physics laboratory, and the best that can be done in the sterile experimental arena pales in comparison to what anyone can see if he or she will only look. Anything—a beach, an open field, a snow-capped mountain—can be used as the starting point for an investigation into the nature of the universe.

The fact of the matter is that everything we see in the world around us operates according to a very small number of basic principles, or laws of nature. There are Newton's three laws that describe the motion of material objects, the three laws of thermodynamics that describe things related to heat, four of Maxwell's equations that describe electricity and magnetism, the single principle of general relativity that describes gravity, and a few more laws (depending on which philosopher of science you talk to) for quantum mechanics, the physics of the sub-atomic world. All in all, then, everything we see must ultimately be connected to no more than fifteen or so laws of nature.

One consequence of this fact is that the universe as seen by a physicist resembles a vast interconnected web. Because every event we see has to be connected to a few general laws, it follows that many events must be connected to each other. Sometimes these connections are mundane, but sometimes they are quite surprising. There is, for example, no obvious reason why white-water in a mountain stream should be related to fluctuations in insect populations or to the concept of the balance of nature, but in chapter 13 this very connection is drawn out in detail.

This universality of the laws of nature is so well known to practicing scientists, so ingrained in us by our training, that we seldom think about it. One of the great rewards I get from writing a book is the chance to rediscover this exciting truth and see it operate in new contexts. After all, the reduction of the complexity of the physical world to a few basic principles surely represents one of the noblest achievements of the human mind. It is important to pause occasionally, lift this ability above the mundane everyday world, and admire it.

For the reader of this book, the existence of natural laws has another important consequence. If all phenomena lead to the same general principles, then you can start with any phenomenon whatsoever and work your way back. If your goal is to understand the universe, a hike through the mountains is as good a place to start as a day at a synchrotron, and a mountain stream can tell you as much as a telescope. That, in a nutshell, is my thesis. In each chapter I begin with something encountered on a walk through the mountains and use it as an entrance point into the universal web. Where the web will lead is anybody's guess, and I wouldn't dream of spoiling your fun by outlining everything in detail; but don't be surprised if you start by thinking about a rock and end up contemplating the universe three minutes after the Big Bang, or start by looking at the swirls in a mountain stream and end by thinking about new ways to move material into space. The universe *is* connected, and this book contains only a few of the examples that illustrate the point.

Finally, let me close this introduction with a comment about an attitude I occasionally encounter. Some people feel uneasy about the idea of taking something as grand and majestic as a mountain and studying it. They feel that by delving into the inner workings of something, we destroy the essential beauty and mystery surrounding it.

From my point of view, nothing could be farther from the truth. When I listen to an opera, I know that there are many levels at which it can be enjoyed. You can just float along with the music, or you can passively follow the story as it unfolds. If you know something of the historical background of the work, so much the better—your appreciation is deepened. If you know enough about music to appreciate some of the finer details of the score or the performance, better still. None of this extra knowledge prevents you from enjoying the opera—far from it. Your enjoyment can only be enhanced and deepened.

I would argue that the same is true of natural phenomena. A rainbow is just as beautiful to someone who understands how it works as to anyone else. The spectacle of a mountain range at sunset is just as impressive to someone who knows how the mountains got to be where they are as it is to the casual observer.

Sometimes it even happens that a deeper knowledge allows you to see things that might otherwise have gone unnoticed. For example, a good deal of my knowledge of the high country comes from the summers I spend in the Beartooth Mountains of Montana. A few years ago, while I was out with a group of friends, I noticed a rather rare phenomenon: a triple rainbow. I quickly gathered everyone together and explained what we were seeing. In the years that have passed since then, every member of that party has made a special effort to tell me how glad they were that I had pointed it out to them. They, at least, didn't appear to feel that extra knowledge had diminished the spectacle one bit.

It is in this spirit, therefore, that I invite you to lace up your boots, pack a lunch, and walk with me along some mountain trails. Maybe I'll be able to point out a few things you hadn't noticed before.

*James Trefil*
*Charlottesville, Virginia*
*1985*

# Meditations at 10,000 Feet

# ONE

# Meditations
# at 10,000 Feet

*Here is no water but only rocks*
*Rocks and no water and the sandy road*

—T. S. ELIOT,
*The Waste Land*

I T HAD BEEN a long, hard climb. We had parked the car at the trailhead when the morning dew was still on the grass, and now, several hours and 5,000 feet in altitude later, we were nearing the end of our hike into the Beartooth range of the Montana Rockies. The trip had not been uneventful. Halfway up we had decided to take a little-used branch of the trail to get away from other hikers. A detour around a swollen stream had caused us to lose the trail completely, and as a result we stumbled onto a beautiful basin nestled high among the rocky peaks. Recovering the trail, we zigzagged up the side of the mountains toward the top. Along the way we were treated to one of the side benefits of mountain hiking— the opportunity to look down on hawks circling lazily in the valley below.

Now, with the last switchback behind us, we were walking up a relatively gentle slope toward our ultimate goal—the tree-less plateau that marks the high country in these mountains.

This part of the trip was less rugged, but the thin air had us gasping for breath by the time we made it to the top. There the magnificent vista spread out before us. To the north and east, the mountains descended to the dry plains of eastern Montana. Due east, the peaks of Wyoming's Big Horn Mountains could just be made out. To the south, the peaks stretched away to Yellowstone Park, surely one of the most beautiful areas left in America. All in all, it was a setting that lent itself to contemplation. As we dropped our packs and settled down to rest in the pale mountain sunlight, I began thinking about the places we had been and the place where we'd ended up.

What exactly does it mean to be at 10,000 feet? Obviously, it means that you have climbed a mountain somewhere, but the question is deeper than that. Why are there mountains at all? Why, of all the possible configurations the earth's surface could have taken, did it turn out to be the way we see it today, with most of the surface below 10,000 feet?

Let me put this question another way. Anyone who has ever flown or driven across the United States knows that most of the land traversed is a level plain whose average altitude is only a few thousand feet above sea level. Only occasionally, as in the area we had climbed that morning, does the land rise above 10,000 feet. This state of affairs holds not only in North America, but everywhere on the earth. For most people, being in the mountains remains a special experience.

This is such an obvious statement that we tend to take it for granted, but it really contains a very deep truth about the way the earth is put together. There is no immediately apparent reason why things have to be this way. In principle, the earth could be a smooth ball with no mountains at all—after all, some of our sister planets have a configuration close to this. The earth could also consist, in principle, of a series of continents whose average elevation was over 10,000 feet, interspersed with oceans. In this case, there would be nothing special about being at a high altitude.

But for some reason the earth is a place where being at 10,000 feet is a special experience. Geologists usually express this fact by showing what they call a hypsometric curve. This is just a graph in which the percentage of the earth's surface above a certain altitude is plotted as a function of altitude. The hypsometric curve for the earth is sketched in figure 1–1.

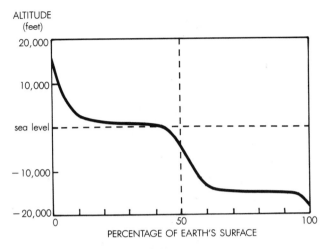

ALTITUDE
(feet)

PERCENTAGE OF EARTH'S SURFACE

**FIGURE 1.1**

Our general impression that mountain heights are relatively rare is borne out by the fact that the graph shows that less than 5 percent of the earth's surface is at 10,000 feet or higher. A mirror image seems to exist at the ocean bottom, where there are very few places much below a depth of 15,000 feet. The data, in fact, show that the land masses on the earth's surface seem to split into two separate types: the level continents a few thousand feet above sea level and the deep ocean floor about 15,000 feet below. Both mountain peaks and ocean trenches are unusual places.

And this brings us back to our original thought. The opportunity to rest and meditate while enjoying the scenery at 10,000 feet depends on the existence of mountains. But why are there mountains at all?

There are actually two aspects to the inquiry. The first is the question of why mountains have formed; the second is the question of why they have survived. Throughout most of human history, most people have simply taken the mountains for granted. They seem to be eternal, and it is natural to think that they have been there since the creation of the earth. The answer to the first question, then, would seem to be that the mountains have simply always been there. In this scheme of things, the second question doesn't arise.

But anyone who has ever spent time in the mountains has

seen features that show that mountains are not eternal. Sitting in the sun on the plateau, I knew that geologists talked about mountains being formed recently. I didn't have a geologist to talk to, however, nor did I have a geology book to consult. Still, I had seen many things on the morning's hike that indicated that the mountains would not be the same in the future as they were that day, nor had they been the same in the past.

The most striking of these indications was a swiftly moving stream that had paralleled the trail for a mile or so. Swollen with melted snow, the water was running high. Rocks the size of bushel baskets were being rolled down the stream bed, and the booming sound that accompanied this motion could be heard for miles. Suppose, just for the sake of argument, that the net effect of that stream was to move ten such rocks per day from the mountainside to the valley below. Of course, no rock would make that journey in a single day, but if you think of the stream as an assembly line, depositing rocks in the valley at one end and picking them up in the high basins at the other, the idea doesn't seem so far-fetched. The stream, then, would move about 10 cubic feet of material off the mountain each day. If we take the flood season to be the months of May and June, then the stream will remove 600 cubic feet of material during that period. Suppose, again for the sake of argument, that during the rest of the year it moves 400 cubic feet, primarily in the form of small pebbles and dust. Then that single stream, in the course of a year, would remove 1000 cubic feet of material from the mountain—enough to fill a moderate-size dump truck.

How many cubic feet of material are there in a mountain? Well, suppose we take the mountain to be about two miles on a side and stand 5000 feet above the surrounding countryside. Then there are roughly $10,000 \times 10,000 \times 5,000$ cubic feet of material in it—about 500 billion in all. This means that the single stream, working year in and year out, would move as much material as there was in the entire mountain in a matter of 500 million years. The lifetime of a mountain, once formed, certainly couldn't be much longer than this. If we take into account the fact that a mountain may have many streams on it, and that there are other processes—rock slides, glaciers, avalanches, etc.—that can wear it down, it is reasonable to expect that a mountain, once formed and exposed to weathering

6

processes, should last no more than a few hundred million years.

This is a rather nice little calculation, because it illustrates an important point about the physical sciences. It is often possible to come up with rough estimates of significant quantities just by making a few simple observations and doing a little mathematical reasoning. Of course, I took the precaution of checking my numbers with a geology book when I got home; but it was in the right ballpark. Mountains do seem to have a life expectancy measured in the hundreds of millions of years.

On the human time scale, this is indistinguishable from eternity. On the scale of the earth's history, however, it is a relatively short time. Four hundred million years is only about 10 percent of the time the earth has been around. Consequently, any mountains that were present at the beginning must have been worn down by the wind and water long ago. This is important, because it tells us that the surface of the earth must undergo continuous change, and that the mountains we see are only the last in a long series of their brethren which have been created and subsequently destroyed over the earth's lifetime.

So, keeping one's eyes open on a mountain hike shows us that there must have been processes in the past that built mountains, and it's not unreasonable to suppose that those processes are going on today. The problem of explaining those forces has played an important role in the history of geology. Mountain building (or, to use the word favored by geologists, orogenesis) was explained in very simple terms by a group of nineteenth-century French geologists, who coined the so-called nappe theory. *Nappe* is the French word for tablecloth, and the theory acquired this name because the building of mountains on the earth was thought of as being analogous to what happens when you push on a tablecloth. You can try this experiment yourself—you'll see that the tablecloth will quickly develop folds and wrinkles as the force is applied. In just the same way, the theorists argued, forces on the earth's surface caused the crust to fold and wrinkle, forming mountain chains in the process. In fact, many of the major mountain chains of the world were formed in this way. The Alps, Himalyas, and the Appalachians are all what geologists call "fold" mountains, and part of the Rocky Mountains is of this type as well.

A tablecloth will not wrinkle all by itself; it has to be pushed.

Sedimentary rocks form in horizontal layers unless they are deformed. These soft rocks have been lifted from the sea bottom, but have not been subjected to any other force. Point Reyes National Seashore, California.

Another example of horizontal sedimentary rocks, this time from the bottom of an old lake. The rocks have been sculpted by the wind. Badlands National Monument, South Dakota.

This sandstone, originally horizontal, was tilted when the Rocky Mountains formed. The openings at the top are old coal mine shafts. Red Lodge, Montana.

This shale, which formed in horizontal layers, has been tilted nearly upright by the forces in the earth's crust. Black Hills, South Dakota.

Another example of tilted sedimentary rocks, this one in the Coronado National Forest, Arizona. Notice the large rock that rolled down the hill—evidence for the continuing breakdown of mountains.

By the same token, the earth's crust must be subjected to an enormous force to cause folding of the solid rock. The evidence for the reality of such forces can be seen in almost any mountain chain, especially in road cuts where the layers of rock, originally horizontal, are now resting at all sorts of angles. The question is how such forces could have been exerted at all.

There has never been any lack of suggestions as to how lateral forces—forces exerted along the surfaces of the earth—could have originated. One poplar theory in the nineteenth century was that the earth was shrinking as it cooled off. In this scheme, the surface of the earth would wrinkle like the skin of an apple that's been left out too long. This theory was discarded by geologists early in this century, but it was a long time dying. I remember seeing pictures of wrinkled apples in my grade school science books in the fifties.

There are actually many arguments against the shrinking earth theory. One is obvious to anyone who has ever driven or flown across the United States. There are only two major mountain chains in North America: the Appalachians on the east coast, and the Rockies and Sierra Nevada along the west. In between are several thousand miles of prairie and high plains. After four and a half billion years, the shrinking apple would surely have its wrinkles distributed more uniformly than that.

Another, more fundamental argument against the shrinking earth theory is that the earth is not, as a matter of fact, cooling off. It's true that there is heat flowing out of the earth's surface and that the interior is warmer than the crust. If there were no way of replenishing the heat lost through this flow, then the earth would indeed have to cool off. In this, it would be no different from the embers of a campfire—if no new wood is added, eventually the embers cool off and stop glowing. In the case of the earth, however, new "wood" is being added to the interior all the time. The decay of radioactive elements in the earth largely balances the heat lost through the surface, so there is very little or no net imbalance. The earth, in other words, is more like a stove whose fuel supply is being constantly replaced than it is like a dying ember. The discovery in the early part of this century of this natural radioactive heating removed a major piece of support for the shrinking earth theory, so we'll have to look elsewhere for the forces that raise mountain chains.

Actually, there is a hint about the solution to the problem in the fact that someone sitting at 10,000 feet is well above most of the earth's surface. If you look carefully at the hypsometric curve on page 5, you will notice that there are two relatively flat regions in it. One of these is the level of the continents at around 2000 feet above sea level; the other is the depth of the ocean floor, about 15,000 feet below the surface of the ocean. Mountain peaks and deep ocean trenches account for only a small percentage of the earth's surface.

From the curve shown on page 12, you see that for all intents and purposes we can treat the surface of the earth as having only two levels. If you ignore the minor deviations from this scheme, you are already starting to think like a physicist. Our new "hypsometric curve," drawn according to this point of view, looks like figure 1–2 on page 12: (Actually, this approximation is not as bad as you might think. If you remember that the continental shelves consist largely of mud washed off the coastal

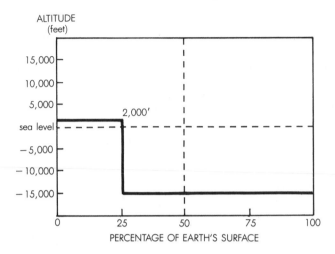

**FIGURE 1.2**

plains by rivers, you will realize that the gradual drop around sea level in the real curve should in fact be much sharper than it is.)

How can we explain this dual nature of the earth's surface? One way to approach the problem is to try to think of other systems that might have this particular feature. Here's one example: Suppose you floated a bunch of wooden blocks on the surface of the water in your bathtub. If you did a "hypsometric curve" for this system, you would find something very much like the one in figure 1–2. There would be only two important levels: the top of the wood and the level of the water. In fact, any system where something light floats on something heavy will exhibit the same behavior. The dual nature of the earth's surface, then, could be explained if we had such light clumps of light material floating on top of some sort of heavier medium.

As it happens, this is the essence of the modern idea of the structure of the earth, the theory of plate tectonics. We'll be talking about this theory in detail in the following chapters. For our present purposes, we need only note that one of its central features is that the continents, composed of relatively lightweight rock, float on a layer of heavier rock in the earth's crust. Thus, there is a close analogy between the blocks of wood floating in your bathtub and the continents floating on the heavier base. In this analogy, the continents, composed of relatively

*12*

lightweight granites, are like the blocks of wood, while the water on which the wood floats is the heavier material (called basalt) that we find on the ocean floor.

Originally, the idea that the continents might not be fixed, but were free to move, was called "continental drift"—a term that you still run across once in a while. The modern theory of plate tectonics is a much more elaborate and comprehensive picture of our earth, as we shall see. Consequently, I'll be using the term "continental drift" only when referring to the historical theory, and "plate tectonics" otherwise.

Once we accept the idea that continents might be mobile, we have a perfect candidate for the lateral force we need to raise mountains. It must sometimes happen that two of the floating continents collide, and when they do, there will be lateral forces exerted—forces that have exactly the properties needed to uplift and fold rocks and bring mountains into existence. So the analogy between wood floating in water and the continents floating on a basalt base would provide just what we need to explain why being at 10,000 feet is such a special experience. High mountains are formed only in those relatively rare situations when the lateral forces associated with the continental motion produce massive uplifts of materials. This simple explanation doesn't prove that the idea of mobile continents is right, of course—we'll leave that task to a later chapter. What it does show is that the idea is plausible. This is no mean feat when you consider we've only used our observations during a mountain hike and a few well-known facts about the earth's surface.

The idea that the continents may not be stationary is a fairly old one, dating back to the early years of this century. The problem has never been to find evidence to support this view, but rather to find some process capable of moving such large land masses. If the surface of the earth is to move, there has to be something generating a force, and that force would have to be huge. Are there any obvious mechanisms that could do the job?

Actually, there is a hint about how continent-moving forces could be generated in one of the things I described on the hike. You don't have to be in the mountains to see hawks circling in the air—this is a pretty universal behavior pattern among birds of prey. You rarely see them flap their wings, their scheme is to find a rising column of warm air and ride it to an altitude of several thousand feet. Then they glide over long distances until they find another column to ride. It is not unusual to see dozens of hawks and vultures circling lazily as they gain height in an area a few hundred feet across.

The basic reason that hawks can fly without flapping their wings is very close to the reason most geophysicists believe the continents move. Both mechanisms depend on the fact that when a material is heated, it expands. Since a given volume of heated material weighs less than its cooler counterpart, it will rise until it either cools off or comes to a region where the surrounding environment is at the same temperature as it is. The physical principle behind the upward motion is just old-fashioned buoyancy—the same thing that makes wood, inner tubes, and (sometimes) human beings float in water.

The actual mechanisms that govern the flight of the hawk and the motion of the continents are slightly different, of course. For one thing, we understand the former pretty well, while the latter remains an area of contention among scientists. For another, the details of how the upward motion takes place are different in the two cases. Let's start with the hawks.

On a sunny day, a layer of air near the ground absorbs heat. Winds move this layer around, and when it comes to a hill or a row of trees or some other obstacle, it starts to rise. Air from the heated layer is sucked into the area that has been vacated by the rising air, so that the net effect of the obstacle is that the air from the heated layer is pulled together to form a single large bubble. The air in this bubble, being heated, is lighter than its surroundings; so it rises. As it starts up, surrounding air moves in to fill the void left by the rising bubble. As the warm air continues its upward course, friction between the rising air and the stationary cooler air outside sets up a pattern of circulation in the bubble like the one sketched below in figure 1–3. Similar effects give rise to the familiar pattern of smoke rings.

At the center of the bubble, the air is rising fast enough to support a bird with outstretched wings or even a manmade glider. All the bird, or pilot, has to do is circle around the central column of air and he will be carried up by the overall motion. This explains why the birds don't have to flap their

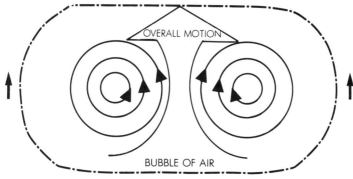

**FIGURE 1.3**

wings and why they always appear to be moving in circles.

One interesting point about the rising air bubbles used by hovering birds is that the temperature difference between the heated and surrounding air doesn't have to be very big to make the system work. If the air in the bubble is as little as five-hundredths of a degree (Fahrenheit) warmer than the surroundings, there will be a buoyant force. Even in the desert, where bubbles can rise through the air at speeds approaching 20 miles per hour, the temperature difference involved is no more than half a degree.

Once a single air bubble has formed near the ground and moved upward, the original obstruction causes the formation of another one. In this way, a stream of hot bubbles will move upward from the obstruction. You can see the same sort of thing in boiling water, where streams of air bubbles rise from a point on the bottom of the pan.

If you watch hawks, you will notice that the circle they describe is at most a few hundred feet across. This gives an estimate of the size of the central core of the bubble, and leads us to suppose that the bubble itself may be only a few thousand feet across. This, in fact, is what meteorologists find when they perform measurements. The rising air bubble, then, is a small-scale response to heating and buoyancy in the atmosphere.

There are larger-scale responses to the same driving forces, responses which go under the name of "convection cells." Convection arises when you have a situation like the one shown in figure 1–4 on page 16, where the air near the surface is heated by contact with the ground but the air above remains cool. In

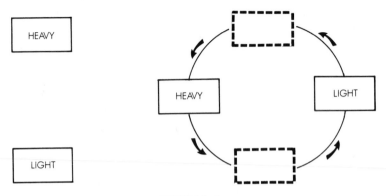

FIGURE 1.4

this situation, the forces of buoyancy will cause the relatively light warm air to rise and the relatively cool air to sink, as shown on the right. Once the warm air gets into the upper position, it will lose its extra heat to the surroundings. Similarly, once the cool air is in contact with the ground, it will heat up. This recreates the original situation, with the warm, light air underneath, so the whole cycle starts again. It's not hard to see that heating and buoyancy will result in a continual turning over of the air, with the net effect being the transfer of heat from the ground into the upper atmosphere. There, the heat will eventually be radiated away into space.

Convections cells can occur on almost any scale. They can be seen around cities and islands, where they are initiated by warm air rising above the heated ground or, in the case of cities, concrete. I once saw a study of a convection cell generated by the large black asphalt parking lot of a shopping mall. On a larger scale, most of our weather is due to the convection cells that form because the polar regions of the earth are cooler than the equator. If the earth didn't rotate, the main flow of the atmosphere would be as shown on the left in figure 1–5 opposite, with warm air rising from the equator and sinking at the poles. In this case, people on the surface would feel the wind blowing from the north to the south.

The fact that the earth does rotate, however, tends to stretch out the simple single convection cell pattern, turning it into the three-celled system shown on the right. Here the central cell, which covers the temperate zones, produces winds that blow

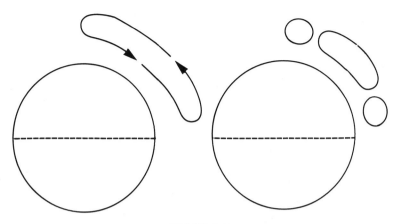

FIGURE 1.5

from the west—the so-called prevailing westerlies. Nearer the equator, on the other hand, the winds blow from the east, producing the trade winds. Between the two zones, where the primary movement of the air is vertical, there is very little wind to move ships along the surface. In the days of sail, this region was known as the "horse latitudes" because ships would sit for weeks on end, use up their feed, and be forced to jettison livestock. The sailors would then see horse carcasses floating in the water. (At least that's the story I've heard.) A similar calm region at the equator was called the doldrums.

The point of this digression is that convection cells can be established in the air over areas that are comparable to the entire surface of the earth. Thus, the forces of buoyancy caused by differential heating can produce structures as small as a few thousand feet (rising bubbles) and as large as many thousands of miles across. Convection cells, therefore, are obvious candidates for our attention when we try to understand how it is that the mountains and the earth beneath our feet can move.

The most widely accepted hypothesis about the origins of continental motion is shown in figure 1–6. The idea is that deep in the earth, heat is being transferred from the hot interior to the outside through the action of convection cells. These cells are located in solid rock, not in any liquid part of the earth. Consequently, the motion is very slow, and the cell takes hundreds of millions of years to complete a single turnover.

*17*

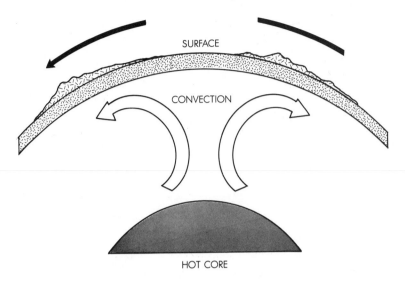

SURFACE

CONVECTION

HOT CORE

**FIGURE 1.6**

This may seem like an unlikely process, but it is the most efficient way for the earth to move its internal heat. Furthermore, we know that "solid rock" can move and be deformed by forces acting over long time periods, so it is not unreasonable to suppose that such convection cells may exist deep within the earth. The mechanism is the same as that for the atmospheric convection cells we talked about earlier. The rock at the bottom is warmer than the rock near the surface, so the force of buoyancy will tend to push the lower rock upward while the cooler rock sinks. The result is the familiar circular motion of the convection cell.

The connection between the interior convection cell and continental motion is simple. The lateral motion at the top of the cell drags the surface along with it, and the continents float along like bits of driftwood on a stream. In this picture, the entire face of the earth, which seems so solid and substantial to us, is nothing more than an incidental byproduct of the process of heat transfer in the body of the planet.

It's a sobering thought. Almost as sobering as the thought that it has only been since the mid-1960s that we have understood the fact that we, our land, and our mountains play such an insignificant role in the scheme of things on earth.

*18*

# TWO

# A Pedestrian's Guide to Plate Tectonics

*"Inestimable stones, unvalued jewels*
*All scattered in the bottom of the sea"*

—WILLIAM SHAKESPEARE,
*Richard III, Act I Scene iv*

D URING THE mid-1960s a major revolution took place in the way we view the world we live on. A new theory called plate tectonics pictured the surface of the earth as a restless and dynamic place, a place that has been changing since the planet formed, is changing today, and will continue to change into the distant future.

Some authors have compared the new view of the geological process to the revolutions brought about by Copernicus and Darwin. I don't agree with this assessment. Both Copernicus and Darwin wrought fundamental changes in our view of the place of mankind in the universe, and plate tectonics has not had nearly the same effect. It is, however, one of those major discoveries—like the discovery of the structure of DNA or the atom—that leads to a burgeoning of knowledge in some important field of study.

Perhaps it would be best to begin our discussion with a def-

*19*

inition of terms. The term "tectonic" is unfamiliar to most readers. It comes from the Greek root *tektonicos*, which originally referred to a skilled builder or carpenter (the same root appears in the word "architect"). The term "plate" refers to the fact that our present understanding of the earth's surface is that it is split up into at least a dozen large, independent plates and many smaller ones. Thus, "plate tectonics" refers to the construction of the earth's surface from plates.

The boundaries of these plates are shown in figure 2–1 on the facing page: Almost the entire continent of North America, for example, is located on a single plate, which begins in the mid-Atlantic and ends on the Pacific coast. The plates themselves are about 100 miles thick and compose what is called the lithosphere of the earth. The plates float on another layer of rock. This layer, called the athenosphere, has a very low rigidity so that it flows easily. The plates, then, float on the athenosphere like a piece of wood on water. The athenosphere is anywhere from 60 to 150 miles thick. Both its existence and its extremely low rigidity have been well documented by the study of seismic waves that have passed through it. This means that there is experimental confirmation of the existence of both the plates and the fluid-like material on which they float.

All of the plates are free to move along the surface of the earth. At any given moment they cover the surface completely, but there is a constantly changing interplay between them. There are many kinds of boundaries between plates. At a converging boundary, two plates which are traveling in opposite directions collide with each other. At a diverging boundary, two plates move away from each other, leaving a void behind. Finally, plates can come together at a neutral boundary and simply move along parallel to each other. Each of these possible plate boundaries produces different geological effects.

It is important to understand at the outset that the plates are not the same as the continents. The Pacific plate, for instance, contains only ocean floor, while the North American and South American plates contain both ocean floor and dry land. The general structure of a plate is sketched in figure 2–2 on the following page. The lower part of the plate is made up of relatively heavy material like basalt, about five times as dense as water. On top of this layer is the lighter continental material.

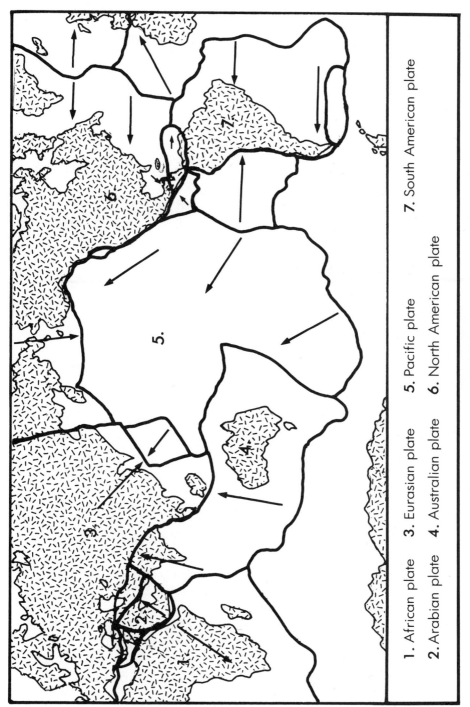

1. African plate    3. Eurasian plate    5. Pacific plate    7. South American plate
2. Arabian plate    4. Australian plate    6. North American plate

**FIGURE 2.1**

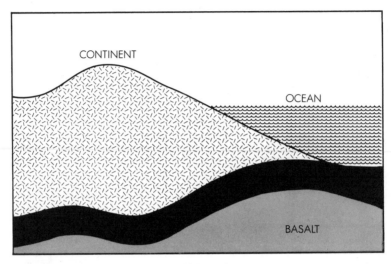

**FIGURE 2.2**

Continents are "light" in the sense that they are made largely of rock like granite, only three times as dense as water. So, while it may seem a little strange to talk of something like granite "floating," it will do so if the material underneath it is heavier, as is the case for the plate shown above.

The best way to think of the continents, then, is to imagine them as bits of debris being carried along on a moving plate. In this scheme, the motion of the continent is merely a reflection of the motion of the plate that is carrying it. If two plates happen to come together at a convergent boundary, and if each plate happens to have some continental material located on that boundary, then the continents will be welded together. The Himalaya Mountains, for example, mark the point where the Australian and Eurasian plates came together. If two plates have a diverging boundary underneath a continent, then the continent will be torn apart as the plates move away from each other. This is happening today in eastern Africa, where the Arabian plate is moving away from the African plate. The Great Rift Valley marks the spot where this process is going on. Finally, if two plates have a neutral boundary and one is moving sideways with respect to the other, the boundary will be marked by geological faults. The most famous neutral boundary is probably the San Andreas Fault in California. The site of many

The lakes in the background mark the location of the San Andreas Fault. The far hills are on the Pacific plate; the photograph was taken form the North American plate. San Mateo County, California.

earthquakes, including the one that devastated San Francisco in 1906, this fault marks the boundary between the North American and Pacific plates.

It is natural to ask how the two-tiered structure of the tectonic plates came into existence. The clearest explanation of the phenomenon requires some knowledge of how the earth was formed some four and a half billion years ago. The best evidence we have indicates that the earth, the other planets, and the sun all formed at the same time from a contracting cloud of interstellar dust. Near the present orbit of the earth, grains of material began collecting into clumps called planetesimals, and the planetesimals began colliding and coming together into the so-called protoplanet. For many revolutions along its orbit, the newborn earth moved through a sea of interplanetary debris, picking up material the way a car windshield picks up bugs on a summer evening. From the vantage point of a hypothetical observer on the surface, this early period in the history of the earth would have been characterized by a rain of meteors falling from the sky. The impacts and the decay of radioactive elements within the planet heated the earth up to the melting point. For several hundred million years, the earth remained molten. During this period a process called differentiation occurred—heavy elements and minerals sank toward the center of the earth, while lighter components rose to the surface. It was during this time that the earth acquired an iron core.

The material of the continents, being the lightest constituents in the solid earth, moved toward the surface during the molten period. Whether all of the present continental material came up at that time or whether additional bits and pieces have surfaced since then due to the action of convection cells remains a matter of debate. What is clear is that early on in the life of the earth all the lightweight material came to the surface, forming the continental crust. No new continental crust is being formed today.

The continents, then, perch on top of the lithospheric plates like people riding on a raft. They move around with the plates and have no independent motion of their own. They are, in the grand scheme of things, little more than flotsam that has risen to the surface of the planet, while the truly important processes are going on inside.

The plates themselves, though denser than the continents, are also floating on top of material that is still more dense. The same process of differentiation which brought the continental granites to the surface also brought the heavier basalt of the lithosphere to its present position. The best way to think of the plates is to imagine a film of oil covering a pot of water that is just starting to boil. The convection in the water will cause the surface to move, and this surface motion will carry the oil film around with it. Pieces of the film will break off into smaller sections analogous to tectonic plates. These "plates" will move with respect to each other, break up into smaller pieces, coalesce into larger ones, and so on. About the only thing you can say about the oil film in this situation is that it will be in a state of constant flux. In just the same way, the lithospheric plates of the earth are constantly changing in response to convection in the interior. Geologists are only now learning to read the history of the plates in the rocks.

When two plates come together at a boundary, several things can happen. One plate may slide below the other, a process

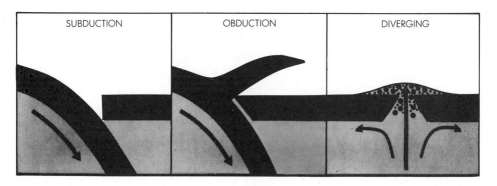

**FIGURE 2.3**

called subduction. This is shown on the left in figure 2–3 above. The subducted plate returns to the earth's interior, where it is melted and ceases to be a part of the earth's outer surface. If there is a continent as a passenger on either of the two plates at a convergent boundary, it can simply float across the subduction zone (if it's on the subducted plate) or be crumpled up to form a mountain chain (if it's on the nonsubducted plate). The Andes Mountains on the western coast of South America, for example, were formed by the forces exerted on the continent by the subduction of the Nazca plate. If there happen to be continents on both sides of a converging boundary, then they will be welded together as both try to float over the subduction zone. The Ural Mountains in central Russia are the remains of the scar that marked the suturing of Europe and Asia. Finally, it sometimes happens that part of the subducted plate is shaved off and lifted to the surface of the upper plate. This process, known as obduction, is shown in the center of figure 2–3 above.

When two plates share a diverging boundary, as shown on the right in figure 2–3, the process is quite different. The separation of the plates leaves a void, and hot magma (liquid rock) from the earth's interior wells up into the vacated space. If the boundary is not covered by continental crust, the result is a mid-ocean ridge. The mountain chain that runs through the center of the Atlantic Ocean—the largest such chain in the world—was formed in this way. If the boundary is covered by a continent, then the continent will be torn apart, and a new ocean will move in to fill the rift thus created, as is going on today in East Africa.

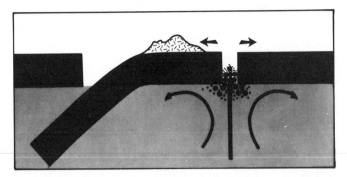

FIGURE 2.4

The motion of the plates, then, is quite easy to understand. As shown in figure 2–4 above, new ocean bottom is made in regions where upwelling materials create a diverging boundary between two plates. As the plates move apart, the molten material fills the space left behind, cools, and is added to the diverging plates. In effect, this process creates new oceanic crust continuously. The other end of the plate may be subducted, as shown. In this case, old crust is being destroyed on one side of the plate at the same time that new crust is being created on the other. Thus, a plate can be thought of as a kind of conveyor belt, bringing material for new crust up at one end and then sending it back into the interior at the subduction zone.

The plate tectonic picture of the earth explains many of the features we observe on our home planet. For example, it explains why there are oceans and dry land, and why so much of the earth's surface is covered with water. If there were no lightweight continental materials on the earth, then the only dry land would be the tips of volcanoes and ocean ridges that result from tectonic activity. There would be no continents as we know them and most of the world would be covered with water. If there had originally been enough lightweight material to cover the plates completely, the end result would have been the same, only this time the sea floor would have been mostly granite instead of basalt. As it happened, however, there was only enough continental material to cover about a quarter of the earth's surface. This means that the earth's surface is two-tiered, with about a quarter of the land located in the upper tier. The ocean basins (the lower tier) fill with water, and although both the shapes and locations of both basins and continents change continuously with time, the 3 to 1 ratio between the two stays fairly constant.

There is another fact that is explained quite neatly by plate tectonics. It is possible to determine the age of rocks in a variety of ways. When these sorts of measurements are made, a clear pattern emerges. Rocks found on continents tend to be very old. In the United States, for example, rocks located between the Mississippi River and the Rockies range from 1 to 2.5 billion years in age. These rocks have existed in their present form for a substantial fraction of the lifetime of the earth. Rocks on the east and west coast tend to be younger, with ages from 200 to 500 million years. But even the youngest of these is older than the oldest rocks on the ocean floors. We know of no ocean rocks older than 130 million years.

This pattern—old continents and young ocean floor—is easy to explain in terms of the motion of plates. The oceanic rocks are created in the process of sea floor spreading at the rate of a few inches per year. At that rate, an area of new rock equal to that of the entire earth's surface could be produced in about 100 million years. Similarly, if we look at subduction zones where ocean floor disappears, we find that 100 million years is about the time it would take for an area equal to the earth's surface to be destroyed.

You can even do a simple calculation to convince yourself that 100 million years should be the average lifetime of oceanic rocks. You know that the North American plate is being created at the mid-Atlantic ridge right now. How long will it be before rocks being added to the plate as you read this have been pushed to the coast of California by the process of sea floor spreading? The distance involved is about 4000 miles, or 250 million inches. At the present rate of a few inches a year, it will take roughly 100 million years to do the job. Since the North American plate is one of the largest on the earth's surface, the fact that the creation of new ocean crust can move material across it in 100 million years establishes the typical lifetime of ocean rocks. This, of course, agrees nicely with what we actually observe.

The reason the continental rocks are so old is also easy to understand if we recall that continents do not undergo the process of subduction. When rocks being created in the mid-Atlantic as you read this are being destroyed 100 million years hence in a subduction zone, the continental rocks on which you are sitting will still be part of some continent. Being so light, they will float over subduction zones, so no melting process will destroy them.

Plate tectonics also provides a simple view of the way that mountains are formed. The collision of continental masses provides the kinds of forces that are necessary to raise mountain chains. This, incidentally, also explains why mountains are usually found at the edge of continents rather than in the center. The fact that the mechanism for forming mountains is simple in principle, however, does not mean that mountains are necessarily simple to analyze. Take one example from real life: a theory about the formation of the Northern Appalachians in the eastern United States.

Up until about 600 million years ago, the land masses that now make up Europe and North America were joined together. Then a divergent boundary opened up beneath the continent, splitting it into two pieces with, perhaps, a few scraps (microcontinents) in between. About 100 million years later, the plates reversed direction and the new ocean began to close. A subduction zone formed in the center. As often happens in such situations, the heat associated with the friction in the subduction zone gave rise to a string of volcanoes—an island arc—in the center of the ocean. The islands of Japan and the Philippines are examples of such island arcs in the modern world. In figure 2–5 below we sketch the appearance of the "Atlantic" ocean as it was 500 million years ago.

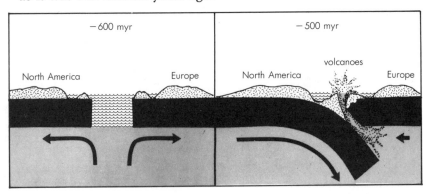

FIGURE 2.5

As the ocean continued to close, the microcontinents were sutured back to the North American continent, and each suturing produced its own (minor) phase of mountain building. Part of the ocean crust was obducted during this process, producing a string of oceanic rocks that are now found in mountains from Georgia to Newfoundland. Around 230 million years ago, the ocean had completely disappeared. The continents of Europe and North American had rejoined, and the shock of the collision is what raised most of the present chain of mountains we call the Appalachians. The materials which were thrust up to form the chain included sedimentary rocks that had formed on the continental shelves as well as volcanic rocks from the (now extinct) island arc. The situation 370 and 200 million years ago is shown in figure 2–6 below.

The second rejoining of Europe and North America did not last long. About 165 million years ago, a new divergent boundary opened up under the continent, but not at the same place as it had done previously. The modern Atlantic Ocean began to form. Europe and North America are still being pushed apart in the modern era by the sea floor spreading at the mid-Atlantic ridge.*

This example should give you some indication of the kind of complexity that can arise when we take the simple ideas of plate tectonics and apply them to situations in the real world. The Appalachians, admittedly, are one of the more complex mountain chains in the world; but because of their proximity to so

*A more detailed description of the formation of the modern Atlantic Ocean is given in my book *The Scientist at the Seashore* (Scribners, 1984).

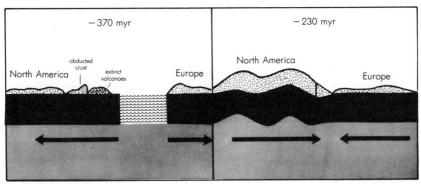

FIGURE 2.6

many major universities, they are also one of the best studied. The detailed history of other mountain chains, such as the Cascade Range in the Pacific Northwest, is not nearly so well known.

From this example we can also draw an important conclusion. If the theory of plate tectonics is correct, then it follows that *there are no permanent geological features on the face of the earth*. Over time scales of hundreds of millions of years, mountain chains are thrown up and worn down, continents separate and rejoin, lithospheric plates move about, break up, split apart. Sea floor spreading starts in one location, stops, starts again somewhere else, and subduction zones do the same. Like patches of oil on boiling water, the plates and their continental passengers present a constantly shifting display as they respond to the forces generated by the internal heat of the earth.

The most amazing thing about plate tectonics as far as I am concerned is not so much its success in explaining the principal features of the earth's geology, but the fact that it is so different from the view of our planet that has prevailed throughout most of recorded history. The idea that seemingly permanent fixtures like mountains and oceans could appear and disappear repeatedly clashes with the commonsense view that the earth is permanent and fixed and has not undergone any major changes since its formation some four and a half billion years ago. That such a shift in outlook could occur at all is a tribute to the ability of the scientific community to give up cherished beliefs when forced to do so by the data. The story of how the new orthodoxy of plate tectonics came to replace the old view of a fixed earth is a fascinating one, as we shall shortly see.

## Problems with Plate Tectonics Theory

Before we embark on a historical survey, however, it would probably be a good idea to pause for a moment and discuss some of the problems that still remain in the theory of plate tectonics. Its successes are impressive indeed; but it would be a mistake to think that there are no unresolved difficulties. In point of fact, the greatest problem with modern plate tectonics is the same one that was raised against the idea of continental drift when it was first introduced. We still don't know exactly how the internal forces operating in the earth make the plates move.

Almost everyone assumes that plate motion has something to do with convective cells in the earth's mantle. Indeed, that is how we introduced the idea in the previous chapter. But if you were to stop and examine the situation a little more closely, you would find that there is very little direct evidence for the presence of convection cells in the hard rocks of the earth's mantle. Indeed, until 1984 there was no evidence at all. Studies of waves transmitted through the earth from distant earthquakes are now begining to show some evidence that the entire mantle, from the top of the liquid iron core to the bottom of the lithospheric plates, is "boiling." This picture, however, is still very tentative and cannot be taken as unalterable.

Given this fact, it should not be too surprising that many of the details of plate motion have yet to be explained. For example, in our discussion of the formation of the Appalachian Mountains, we saw that plate motion is an erratic, stop-and-start kind of thing. How does this sort of behavior arise from convection cells? We don't know. There are plenty of ideas and theories floating around, but none that is capable of convincing the majority of geophysicists that it is the final answer.

You can get some flavor of the theoretical debate going on today by looking at the three models in figure 2–7 below. On the left we have the conventional picture of plate tectonics, with each convection cell controlling a single plate. Alternative models are shown in the center and on the right-hand side. In the central model, many smaller convection cells operate beneath each plate.In the model on the right, the cells operate only at the edges of the plates, with none under the main part of the plates themselves. With smaller cells, heated material doesn't have to move so far

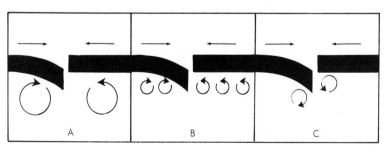

FIGURE 2.7

to complete its cycle. In addition, with many small cells competing with each other, the erratic nature of plate motion becomes easier to understand—it results from changes in a few of the small convection cells and consequently does not require the large changes that would be necessary to reverse the motion of the large cells shown on the left.

On the other hand, if there really are a large number of convection cells under North America, it's hard to understand why the continent stays together. Why doesn't the plate split up onto a large number of smaller plates? Also, a few small cells at the plate boundaries would have to produce an enormous amount of energy to pull the plate along. Where does this energy come from? At present, there is no satisfactory answer to this question.

In sum, the theory of plate tectonics provides an extremely useful and simple picture of the geological processes going on at the earth's surface. There seems to be little doubt that the fundamental process is the motion of plates floating on the athenosphere. There are, however, some essential questions about how such motion arises that must be answered before we can claim to have a complete understanding of our home planet.

## The Proof of the Pudding

Sometimes the most important advances in science come from unexpected quarters. For example, if you had to guess where the critical evidence in support of plate tectonics comes from, your first thoughts might be of classical geology—the study of mountain ranges and other large-scale formations. In fact, the crucial evidence for the theory doesn't come from classical geology at all, but from the seemingly unrelated field of magnetism, and specifically, the study of the magnetic properties of rocks.

Anyone who has used a compass is aware that the earth possesses a magnetic field. The origin and detailed properties of the field needn't concern us here; all we have to know is that the earth's magnetic properties cause any small magnet (like your compass needle) to point toward the north magnetic pole (which, at the moment, is somewhere in Canada). In fact, any magnet will line up in a north–south direction if it is left free to do so.

The first thing you need to know to understand the line of evidence that led to the confirmation of plate tectonics is that atoms and small groups of atoms can act as magnets, and consequently can be thought of as analogous to a compass needle.* In a naturally occurring iron ore like magnetite (or "lodestone"), the forces between atoms in the ore cause more of the atomic magnets to point one way than the other, so that the entire lump of ore acts like a small magnet. The Chinese used this fact to construct the first compass over a thousand years ago. They floated a chunk of magnetite on a cork in a pan of water, where the rock could twist around (taking the cork with it) until it pointed to the north magnetic pole.

The interatomic forces that turn a piece of iron ore into a magnet operate by causing individual atoms to line up like soldiers on parade. So long as the temperature of the material is below a certain value, the force is strong enough to keep the atoms aligned even though the heat causes them to jiggle around a bit. If the temperature exceeds a critical value, however, the motion of the atoms becomes too great for the force to control and the alignment disappears. This is why heating a magnet destroys its magnetic properties. The critical temperature at which the material ceases to be a magnet is known as the Curie temperature.

To see the implications of this situation for geology, just think of what happens when hot material from the earth's interior comes to the surface. If the rock is molten, then all of the atoms in it are free to move around. Such a rock will have no net magnetism. Suppose, for the sake of argument, that the rock contains atoms which will eventually combine to form magnetite. When the material cools, the rock will crystallize, and in this process tiny grains of magnetite will be formed as part of the overall structure. This process takes place well above the Curie point, so the atoms, though locked into the lattice of the newly formed rock, still have no net magnetism. As the cooling progresses, the Curie point is reached and the atoms in the

---

*A more complete discussion of the nature of atomic magnetism is given in my book *The Unexpected Vista* (Scribners, 1983).

magnetite begin to align themselves. Because this process occurs in the earth's magnetic field, the atomic magnets will line up pointing toward the north pole. Provided the rock is not disturbed or heated again, the grains of magnetite will "remember" the direction the earth's magnetic field had when they cooled. The study of the magnetic memory in the earth's rocks is called paleomagnetism.

During the nineteenth century, studies of the remnant magnetism in recent lava flows worldwide showed that they were all formed in a magnetic field essentially the same as the one which exists today. By the early 1900s, however, some puzzling data began to come in. Large formations were found in which the rock was magnetized, but in a direction *opposite* to the present magnetic field. In other words, the magnets pointed toward a "north" pole somewhere in Anarctica. For a long time, geologists tried to blame these anomalies on some sort of chemical or structural change that took place in the rock after it was formed. Indeed, a few minerals do have a property such that their magnetic poles reverse spontaneously in time. There weren't nearly enough such examples to explain the accumulating data, however; so by the early 1960s it was becoming obvious that the magnetic record of the rocks was telling us something very important about the earth itself.

As it became possible to determine the dates of rocks by the techniques of radioactive dating (see chapter 7), the worldwide data on normal and reversed magnetism began to show a single overall pattern. Despite the fact that it seemed impossible, the data clearly showed that the earth's magnetic field had reversed itself not once, but many times in the past. The type of data used to establish this fact is shown on the left in figure 2–8 below. At a given site, either on the sea floor or on a continent,

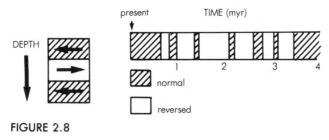

FIGURE 2.8

succeeding layers of rock show a remnant magnetism that alternates between a normal and a reversed direction. By comparing many such sites, a time reversal scale like the one shown on the right can be constructed. We live today in a period arbitrarily defined as having a "normal" magnetic field—a period which has lasted about 700,000 years so far. Before that, there was a long epoch of "reversed" magnetic field extending back to about 2.9 million years ago. This reversed epoch was punctuated by short periods (called "events") during which the earth's field was again normal. The field appears to complete its switch (that is, create an "event") in a relatively short time—perhaps a matter of only a few thousand years.

Although the problem of explaining why these reversals take place remains one of the major unsolved questions in geophysics, the fact that they occur cannot be contested. At least 150 such reversals are known to have taken place in the last 70 million years. The strange behavior of the earth's magnetic field constitutes one piece of the picture which eventually led to the acceptance of plate tectonics.

About the time the magnetic reversal scale was being established in 1965–66, a seemingly unrelated set of data were recorded in the Pacific Ocean off the coast of South America. The National Science Foundation research vessel *Eltanin* was undertaking a systematic survey of the ocean floor. Part of the survey consisted of towing a sensitive device for measuring magnetic fields above the ocean floor. From the readout of the device and the known properties of the earth's magnetic field, it was possible to determine the remnant magnetism of the rocks on the ocean floor. On one data run, known as *Eltanin 19*, the ship crossed the East Pacific Rise, an underwater mountain chain at the joining of the Pacific and Nazca plates. The ship obtained magnetic measurements on both sides of the ridge, and the data is shown in figure 2–9 on page 36.

Scientists at the Lamont-Doherty Geophysical Observatory at Columbia University who saw these results in early 1966 noticed the obvious similarity between the reversals of remnant magnetism in the rocks and the reversals of the earth's magnetic field. This is illustrated by the reversal time scale under the *Eltanin 19* data. The explanation of the data that finally emerged was simplicity itself. If a mid-ocean ridge marks a diverging

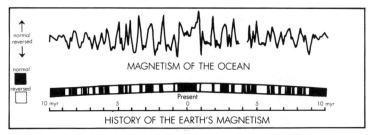

**FIGURE 2.9**

boundary between plates, then the upwelling rock will be hot—
well above the Curie temperature. As the rock moves to the
surface, it will cool, and the present direction of the earth's
magnetic field will be imprinted on it. The first "reversed"
stripe, then, must have been laid down during the most recent
reversed period, over 600,000 years ago. It is now about twenty
miles from the ridge because the plates have spread apart by
that much in the intervening period, and the new rock which
has risen to fill the void has responded to the present, normal
direction of the magnetic field. The symmetrical stripes on the
ocean floor, then, are simply the record in rock of the joint
processes of sea floor spreading due to tectonic activity and the
reversals of the earth's magnetic field.

After the *Eltanin 19* data was announced, evidence to support
sea floor spreading accumulated at a rapid rate. Striped patterns
were observed at the Reykjanes Ridge near Iceland, at the Juan
de Fuca and Gorda ridges off the coast of Washington State,
and in many other places in the Pacific and Atlantic Oceans.
Once people knew what to look for, the symmetric patterns
began appearing everywhere. The full picture of the Juan de
Fuca Ridge off the coast of Washington, for example, is shown
in figure 2–10.

As evidence for sea floor spreading was accumulating in laboratories around the world, researchers at Columbia University
uncovered another important piece of the plate tectonic puzzle.
The Lamont-Doherty Observatory had collected a large number
of deep sea cores—long vertical columns of rock and mud taken
from the ocean floors by research ships. The material in these
cores had drifted down to the ocean floor over long periods of
time, with the oldest materials being at the bottom of the core.
Since some of the debris on the ocean floor comes from con-

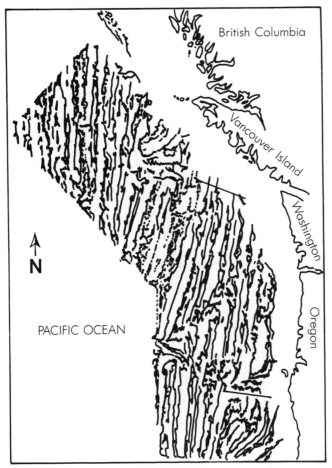

FIGURE 2.10

tinental dirt washed into the sea, there were occasional small grains of magnetite mixed in with the other materials in the cores. During their descent through the water, these grains would be free to orient themselves in a north–south direction, but once they were incorporated into the ocean floor and packed down, they would be immobilized. Consequently, the magnetite grains in the cores taken up by a research ship would "remember" the earth's magnetic field.

A sample core is shown above. The direction of the magnetization of the material shows a characteristic alternating pattern, and all the changes in the earth's magnetic field seen in figure 2–11 can be observed in the material in the cores. Thus,

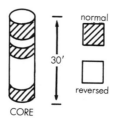

**FIGURE 2.11**

studies of deep sea cores provide a third independent confirmation of the changes in the reversals of the earth's magnetic poles.

If we add this to the data from remnant magnetization of rocks whose age has been determined by radioactive dating and the evidence for sea floor spreading produced by *Eltanin 19* and later work, there can be no question that the earth's magnetic field suffers irregular reversals, and that the record of these reversals is contained in rocks at the surface. Once this fact is established, the symmetrical striped patterns of magnetization at the mid-ocean ridges provide incontrovertible evidence for the motion of the lithosperic plates. Thus, it is fair to say that since the mid-1960s tectonic motion on the surface has been an established fact of earth history.

# THREE

# In the Matter of Alfred J. Wegener

*"Plots have I laid, inductions dangerous..."*

—WILLIAM SHAKESPEARE,
*Richard III, Act I Scene i*

A TYPE OF folklore has grown up around certain figures in the history of science. Men like Copernicus, Galileo, and Wegener are said to have been prophets crying in the wilderness, ignored by their colleagues, perhaps even persecuted for their ideas. They are often cited by proponents of fringe views in science—UFO enthusiasts, psychics, flat earthers—as examples of the close-mindedness of the community of conventional scientists.

As with most folk tales, it is not always clear what lessons one can fairly draw from the facts. The works of Copernicus were widely taught at European universities both before and after his death without any suggestion of persecution either by the Church or his fellow scholars. On the contrary, he occupied a position of some authority in Poland and served on a number of royal commissions. Had Galileo been a little more interested in his science and a little less interested in provoking a confrontation with the papal authorities, he could probably have lived out his years without ever being threatened with a trial for heresy.

The case of Alfred J. Wegener is a bit more complex. During the early years of this century he proposed a theory which held that the continents were not stationary, but moved. This came to be known as the theory of continental drift, although scholars tell me that the term "continental displacement" would probably be a better translation of Wegener's original German. According to the legend that has grown up around him, Wegener's theory was rejected out of hand, despite the presence of the evidence in favor of it, and he was shunned by the scientific community for opposing the orthodoxy of his day. Now that we have an understanding of the theory of plate tectonics (which incorporates Wegener's main thesis), it would probably be a good idea to look at the history of how his ideas went from disfavor to acceptance. In the process, we'll learn something about the way that science works and learn how to evaluate claims of martyrdom from people who currently find themselves outside the pale of conventional science.

First, some biographical facts. Alfred J. Wegener was born in Berlin in 1880 and took a doctorate in astronomy. His main scientific interest was in meteorology, but he also had a keen interest in exploring the Arctic, particularly Greenland. In this, he seems to have shared a passion of his time—one that, I must confess, has always been a mystery to me. Winter is nasty enough without seeking it out, as far as I'm concerned. In any case, Wegener participated in several expeditions to Greenland. After World War I, in which he served as an officer, he joined the Meteorological Research Department of the Marine Observatory in Hamburg, of which he eventually became director. He died in 1930 while leading an expedition to Greenland. All in all, this is not the biography of a man who has been shunned by his colleagues, nor does it show a lonely prophet crying in the wilderness. Wegener was, in fact, a man of substantial achievements in his own field of work, and was widely recognized as such by his colleagues.

It was not as a meteorologist, however, that he made the reputation that he has today. He apparently was interested in geology, and in 1915 he published *The Origin of Continents and Oceans.* In this book he marshaled the evidence for his theory that the continents were once all joined together in a great land mass he called Pangaea ("All Land"). According to Wegener, this land mass split into the present continents millions of years ago; the tides in the earth provided the driving force to move the continents. At the time of the publication of Wegener's book, the most widely accepted theory of the earth's geological history was that the earth was cooling off and shrinking, and that it was this shrinking motion that was responsible for all features of the surface.

Despite the fact that Wegener was an outsider to the geological community and despite the fact that his theory contravened the conventional wisdom of the day, it was not ignored. It was widely discussed in Germany—criticized at Frankfurt, treated sympathetically at Marburg. In general, geologists in Germany seemed to oppose it, while geophysicists seemed to support it. In 1921, for example, Wegener said he didn't know a single geophysicist who was opposed to his idea.

In 1922, however, Wegener's book was translated from German into English and a number of other languages, and his troubles began. Harold Jeffreys, the dean of the British geophysical establishment, pointed out that if there were tides in the earth's crust strong enough to move the continents, then they would also collapse mountain ranges, make the sea floor perfectly flat, and stop the rotation of the earth within a year. At about the same time, American geologists, who would later form the main cores of resistance to continental mobility, brought out pointed and detailed criticisms of many of Wegener's geological arguments. In the face of the theoretical arguments of Jeffreys and the evidence amassed by the American field geologists, the theory of continental drift was pretty well abandoned by 1930. The important point to note here is that it was not ignored or rejected out of hand. *Based on the information available in 1930*, it had simply been weighed in the balance and found wanting.

This is an important point. With 20–20 hindsight, it is easy to criticize scientists in the 1930s and 40s for ignoring and rejecting a theory whose main tenet later turned out to be right. To do so, however, is to commit a grave injustice to a group of hardworking and fair-minded people. The real question we have to ask is this: considering what was known at the time, was their decision the right one to make? Was it, in other words, consistent with the scientific method as it ought to be practiced? I would argue that the answer to this question is yes, and I think the reasons I've come to that conclusion constitute a lesson in the way that scientific change takes place.

Let's go back and look at the evidence Wegener presented to support his hypothesis in a little more detail. Basically, it can be broken down into five categories: (i) geographical, (ii) geological, (iii) evidence from fossils, (iv) evidence that the poles of the earth had moved, and (v) geodetic evidence that Greenland was moving away from Europe at a measurable rate.

The first category of evidence is obvious to anyone who has looked at a map. As early as 1620, Francis Bacon pointed out that the coastlines of the Americas and Europe-Africa seemed to fit together like pieces of a puzzle. Wegener improved the argument by showing that the edges of the continental shelves, the real borders of the continents, match even better than the coastlines. This fact would be easily explained if the continents had once been joined; indeed, that is precisely how it is explained in the modern theory of plate tectonics.

The geological evidence centered on two points. One involved some formations of very old rocks. The largest part of these formations were in Africa, but there were a few similar formations in South America located at about the points that the two continents would have fitted together had they once been joined. Wegener argued that these formations had once bridged the continental boundaries and were subsequently separated by continental motion. The other point involved the evidence for the twofold nature of elevations on the surface of the earth—evidence we talked about in chapter 1.

The fossil evidence supporting Wegener's theory centered on a few remains of animals who were found only in Brazil and South America, and nowhere else—for example, the Mesosaurus, a small reptile. The argument was that such animals must have developed at a single location. Finding fossils on two continents, then, would be evidence that those two continents had been joined together when the species in question came into existence, and that the separation we now see must have occurred sometime after that event.

42

As a meteorologist, Wegener was particularly interested in evidence relating to the past climate of the earth. He knew, for example, that the massive coal deposits of North America and China indicate that those places once had a tropical climate— a climate they do not have today. He argued that this showed that both areas were once closer to the equator than they are now. In a similar vein, he attempted to locate the position of the south pole by recording the location of the piles of rocks ("till") that mark the furthest advance of glaciers. You could do a similar calculation for the last Ice Age, for example, by marking out the places in America, Canada, and Europe that indicated the farthest advance of the ice caps and then put the pole at the center of the circle you had determined in this way. Wegener argued that his analysis showed that the south pole had moved since ancient times.

Finally, Wegener brought in a bit of direct evidence. There were some measurements that seemed to indicate that the distance between Greenland and Europe was increasing. This evidence depended on comparing two different measurements of this distance taken at different times, and each measurement was questionable in itself. Determining this sort of distance to within a few inches was not really within the capability of the early nineteenth-century scientist, so that comparison with later measurements was not reliable. Nevertheless, if you just took the numbers at face value, they seemed to support Wegener's thesis.

So there you are. The case has been presented, and you are a geologist in 1930 trying to decide whether to adopt continental drift or stay with a less radical theory. What would be going through your mind?

Perhaps the best way to understand what might happen is to follow along as I try to analyze the evidence. I am fairly conservative by nature; if you rated physicists on a scale of 1 to 10, with 10 being the most conservative, I'd probably rate a 6 or 7. My thoughts, then, might well be typical of what scientists do when confronted with this sort of situation.

Let's start with the direct evidence concerning Greenland. To prove that the island is moving away from Europe, you have to make two measurements and compare them. Each measurement will have some built-in error associated with the equipment and technique being used, and it is imperative that the error be smaller than the effect you are trying to measure. It would do you no good to try to compare two sticks whose lengths differed by less than a thousandth of an inch if the only measuring instrument you had was an ordinary yardstick. Any difference between the sticks would be buried in the uncertainty of the measurement itself.

Suppose you believed that Greenland was moving away from Europe at the rate of 10 feet per year (this is much larger than the value we expect today). If you take two measurements a year apart, you would have to do each one to an accuracy of better than 5 feet if you wanted to make a meaningful measurement. To see what I mean by this, look at the graph in figure 3–1. In this graph we show two measurements, and the vertical lines bracketing each point represent the margin of error associated with the data. You can think of the line as saying "the real distance could be anywhere between the top and the bottom." In the figure, we have assumed that the error in each measurement is 20 plus or minus feet. Now the actual number that measures drift is the difference between the data points which lie at the center of each line. In the graph it is 10 feet (60 minus 50). But this does not mean that you can conclude from this measurement that the actual distance has changed by 10 feet. The error bars indicate that while the first distance is probably 50 feet, it could be as high as 70 or as low as 30. Similarly, the second measurement could be as high as 80 or as low as 40 feet. Thus, all we can tell from this graph is that the distance could have shortened by 30 feet (40 minus 70), lengthened by 50 (80 minus 30), or have changed by any value in between, *including zero.*

The data that Wegener used to argue for continental drift contained this wide margin of error: the possible errors in the measurement were larger than the claimed drift of Greenland. Consequently, scientists tended to discount the evidence, as well they should have. Their skepticism has been borne out by modern, more accurate measurements, which show none of the movement that Wegener claimed was there. So as far as direct proof of continental mobility is concerned, a reasonably skeptical observer would have to say that Wegener's geodetic argument was invalid.

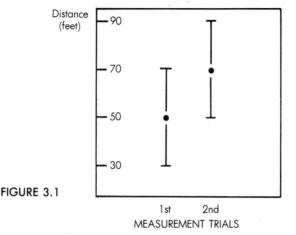

FIGURE 3.1

What about the fossil data? Well, the favorite theory of the time, the shrinking earth, postulated the existence of land bridges, now sunken, linking the continents. We know that at least one species of animal—human beings—migrated from Asia to North America across the Bering Straits at a time when there was still solid ground there. If the first Americans could do it, why not the Mesosaurus? There is no problem in incorporating the fossil record into the shrinking earth theory.

To be fair, the shrinking earth theory has some problem accounting for the disappearance of the land bridges. If they were made of continental material, you have to explain why they sank into the heavier basalt of the ocean floor. If they were originally basalt, then how did they get above sea level in the first place?

What about the geological evidence? Here the picture is a little murkier. It's true that there are a few rock formations which seem to continue from Africa to South America, but how much weight should this fact be given? The fact that two formations are similar today does not necessarily mean that they were formed at the same time in the same place. Minerals change continuously in response to the conditions around them; because two minerals are the same now does not necessarily mean that they have been the same in the past or will be the same in the future.

Besides, of all the different geological formations that exist on the two sides of the Atlantic, it would be amazing if a few didn't match up. The continuation of the few formations that Wegener cited may have been no more than the results of chance. In this sense it would be similar to someone drawing a full house in poker—the odds are against it, but given enough poker hands, it's sure to happen sooner or later.

As far as the elevations on the earth's surface are concerned, this can be explained without continental drift. If the earth really went through a molten stage, then you would expect the lightest material to float to the top. As we saw in chapter 1, this type of floating system is all that is needed to explain the hypsometric curve. There is no need to suppose that once the crust hardened, there was further lateral movement of the continents. All you need is to have continental material rise to the top and collect there during the initial molten phase.

The evidence for the movement of the poles is a bit harder to evaluate. The argument here sank quickly into a morass of technicalities. Field geologists disputed Wegener's claim that the formations he identified as glacial tills were caused by glaciers at all. They also disputed the claim that the Pennsylvania coal fields originated in a tropical climate. On the evidence available in 1930, you'd probably have to say that the question of a moving pole was very much open. It is a hard question to resolve, as is shown by the fact that it was still being debated in the 1960s.

As to the final piece of evidence for Wegener's theory, the jigsaw fit of the various continents, it must be admitted that it was explained by Wegener's hypothesis. It can, however, be explained in other ways as well. For example, one theory that had some currency until the 1960s held that the earth had been compressed during the early gravitational collapse and was now rebounding, much as a rubber ball will expand after being squeezed. In an expanding earth, the continents could be thought of as a thick layer of paste covering the surface of a balloon. As the balloon is blown up, the paste will split apart. If you look at the paste at a later time, the pieces will indeed fit together, but they will not drift. Thus, it was possible to explain the facts without Wegener's strongest argument and hence without accepting his main thesis.

When you tally up the worth of the arguments for continental drift put forward in Wegener's book, then, you find that of the five pieces of evidence, one (the motion of Greenland) can be rejected out of hand. The data is much too questionable to be of any use in the argument. Of the remaining four points, the geological evidence can be accommodated within the shrinking earth model, as can the fossil evidence if we can find some way to make the hypothetical land bridges sink. The evidence for a moving pole, as we have seen, was at best controversial, and the only thing that could be done on this point in 1930 was to adopt a wait-and-see attitude until the experts sorted things out for themselves. As for the mutual fit of the continents, there was no obvious way to explain this by reference to the shrinking earth.

This sort of ambiguous situation is typical of what scientists have to use as a basis for decision every day. Perhaps this explains why they tend to hesitate to make statements. But if every decision were as simple as it sometimes appears in hindsight, there would hardly be any need for scientists to devote years of education and training to learning their craft.

So, if you were a geologist in 1930, you would have to choose between Wegener's theory, which had a simple explanation of the fitting of continents but no plausible mechanism for making continents move, and the shrinking earth theory. If you chose the latter, you would have to assume that the problem of the land bridges was one of those details that could be worked out later. And as far as the fit of the continents goes, it would probably not seem an urgent problem. Solving it might be interesting, of course, but it could be thought of as another detail whose solution would become clear once the general problems of the earth's structure had been solved.

If, on the other hand, you were to take continental drift seriously, you would be pinning all your faith a frail hope. In return for explaining the fit of the continents, you would have to accept a theory that not only flew in the face of everything you had been taught but postulated a motion of the continents for which no reasonable mechanism could be suggested. There's little question in my mind about how I would have reacted to this choice—I would have ignored the continental drift theory and gone on with whatever I was doing at the time. Any other choice would have been foolish.

I probably wouldn't have gotten upset about continental drift the way some orthodox geologists did, but that may be because my background is in physics and not the earth sciences. It has always been my impression that debate in paleontology and related areas is a good deal more vituperative than what goes on in my own field. For example, not many years ago over a dinner table, a geologist called one of his fellows, a solid and conservative scientist who had defended the theory that dinosaurs had been wiped out by an asteroid hitting the earth, a "fraud and a charlatan." I've never seen a public display of peevish bad manners like this between physicists. Like members of the House of Lords in England, physicists tend to couch their most vicious thrusts in impeccably polite phrases. If you hear a physicist saying: "I don't understand how you come to that conclusion," the real meaning is probably something like: "Look, stupid, that whole argument is pure baloney."

In any case, I hope I have convinced you that the refusal of geologists to take Wegener seriously in the 1920s was not only thoroughly justified but was really the only choice an intelligent person could make at the time. In hindsight, we know that Wegener was more right than wrong, but that's only because we have more information now than we had then.

## But Wasn't He Right?

After all this discussion of why a scientist in the 1930s shouldn't have believed Wegener's theory, what are we to make of the fact that the continents actually do move? Doesn't this triumph show that the arguments I've presented must be flawed in some way?

There are two important points that can be made on this question. First, Wegener did not propound the modern theory of plate tectonics. He had, for example, no notion whatsoever of lithospheric plates. Scientists in the 1930s, therefore, did not reject a correct theory, but simply rejected a theory in which one element—the movement of continents—happened to coincide with one element of the theory that later turned out to be correct. Other aspects of continental drift turned out to be completely wrong. For example, Wegener argued that the motion of the continents would cause the average elevation of dry land to increase—something that is not a part of modern plate tectonics. Consequently, continental drift, while it may have prefigured modern theory, was not in itself a workable solution to problems in geology.

In addition, geologists did not have the luxury of considering continental drift in isolation. As is usually the case in frontier areas, there was not one, but a multitude of theories clamoring for approval. In retrospect, it's always easy to pick out the ones that have points in common with what later turns out to be the correct point of view. We remember a single idea like continental drift and forget all the wrong guesses. There's nothing wrong with this tendency; it's part of human nature. Without it psychics and palm readers would probably go out of business, since their stock in trade is to make so many predictions that at least one has to be right. Maybe the best way to make this point is to put continental drift in context by giving a brief description of a few of the other theories that were around a century or so ago.

One, more outrageous than Wegener's, was proposed by the famous French geologist Léonce Elie de Beaumont. From his geological studies and some mathematics, he was convinced that the appearance of mountain chains on the earth was linked to pentagons drawn on the earth. A map of Europe with its appropriate pentagons is shown in figure 3–2 on page 50. Even though Beaumont was a major figure in European geology, his theories were given a less enthusiastic reception than Wegener's.

A more serious attempt to deal with the formation of the earth was given in 1907 by the American astronomer William Pickering. Pickering proposed a model in which the earth was formed already differentiated, with the heavy iron core forming first, followed by succeeding layers of ever-decreasing density. According to Pickering, the earth began with an unbroken solid crust of roughly the same composition as the continents. Then, in a cataclysmic event that took place while much of the terrestrial material was still molten, three-quarters of the outer mantle was torn off to form the moon. (If the early earth were spinning rapidly enough, there would be no difficulty in finding forces large enough to accomplish this feat.) The birth scar of the moon was the Pacific basin, and the shock of formation caused the remaining quarter of the continental crust to shatter

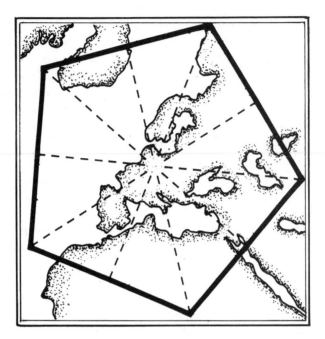

FIGURE 3.2

and fly apart. The present continents drifted through the still molten earth until they were slowed down by friction. Once stopped, the earth's cooling locked them into place, where they are now found.

Note that in this theory the continents moved only once, very early in history. Pickering's theory also explains every geological fact that Wegener's does—the continuous geological formations and the way the continents fit together. And although Pickering's theory doesn't explain the fossil data, Wegener's couldn't explain why the moon is so much less dense than the earth—roughly equal odds for the two theories.

The point of this excursion into history is to do something that is seldom done in writing about science. We tend to concentrate our attention on successes and to judge theories of the past on the basis of how closely they resemble theories of the present, a process that seriously distorts our view of the way things really happened. There is, after all, a lot to be learned by considering the losers as well as the winners.

## The Turning Point

Thinking about the fate of the theory of continental drift leads us naturally to the more general problem of how the scientific community decides when to accept one thesis rather than another. The first point that has to be made is that there is no ruling body in science charged with defending orthodoxy, no Pope whose word is law. Individual scientists go through the kind of reasoning we've outlined above when they analyze problems, and the opinion of "Science" is nothing more than the consensus of many individual scientists.

In questions like those surrounding continental drift, the main participants in the decision are likely to be scientists working at universities, doing what is called basic research. University scientists have many duties other than research: they are responsible for teaching undergraduates, training graduate students, and helping to administer the university. Consequently, their most precious commodity is research time. When such a person looks at something like continental drift, he doesn't think only in abstract terms. The real question is not "Is this right?", but "Is it worth devoting some of my time to this idea?"

Look at it this way. A scientist's working life usually doesn't start until his late twenties and ends in his sixties. It spans about forty years in all. Any major research effort is likely to consume at least five years. Indeed, moving into a new field of research is likely to require a year just to learn the techniques, build the equipment, and do the necessary background studies. A five-year research effort, furthermore, will consume around 12 percent of the total time available in the researcher's life. So when a scientist looks at something like continental drift, he has to ask: "Is the probability of a payoff from research in this field high enough for me to risk twelve percent of my career by going into it?" If enough scientists answer this question in the affirmative, then the field will grow and attract new followers. A bandwagon effect may even set in as new workers solve outstanding difficulties and the odds of success increase. If, on the other hand, enough people answer in the negative, the field dies of neglect and its problems never get resolved. Clearly, a geologist in 1930, looking at Wegener's proposal, would have been taking much too high a risk had he plunged into working on it. Scientists therefore stayed away in droves and, in retrospect, it turned out they were right. It would have been thirty-five years before anyone who had sided with Wegener could have seen the light at the end of the tunnel—almost an entire professional lifetime.

When, then, did it become reasonable for scientists to start taking mobile continents seriously? As with anything that involves human behavior, it's not possible to give a clear answer to this question. There are at least two factors that scientists take into account in making this sort of judgment, factors that I will label "experiment" and "theory." Actual judgments are made by mixing these two together, and each scientist has his or her own way of weighing the two in order to come to a decision. The key question under "experiment" is whether or not the effect to be explained has, in fact, been firmly established. The key question under "theory" is whether the new idea can either be fitted coherently into known physical principles or carries with it a new set of principles that can be verified.

Each major change in scientific thought presents the scientists of the time with a different mix of experiment and theory. General relativity, for example, was so compelling a theory that scientists were willing to accept it with a minimum of experimental verification. Plate tectonics, on the other hand, was accepted because of the overwhelming evidence for sea floor spreading discussed in the last chapter. Its theoretical underpinnings were weak; indeed, the problem of what causes the motion of the plates has still not been satisfactorily resolved. These two examples show extreme cases of the acceptance of new ideas and illustrate the complex nature of scientific decision making.

Each scientist, watching the changing interplay of theory and fact that accompanies research, constantly updates his or her judgments about new ideas as time goes by. I have argued that it was reasonable to reject continental drift in 1930. It is just as reasonable to accept plate tectonics today. Where was the crossover point? I suspect that the firm establishment of sea floor spreading in the mid-1960s marked the point at which the majority of earth scientists were nudged over the boundary between belief and skepticism on this issue. If you will again take my own thoughts as being typical of scientists at large, we can look at this process in some detail.

When the paleomagnetic evidence began to show clear proof that the earth's magnetic field had reversed repeatedly in the past, I would have started to have doubts about the proposition that the continents had been fixed in the past. These doubts wouldn't have made me think about changing my research efforts, but they would have prepared the ground for what came later. When the *Eltanin 19* data showed a precise correspondence between the magnetic stripes near an ocean ridge and the magnetic timetable, I would have moved the whole issue to the "open question" category and begun to think seriously about accepting continental drift. When the second sea floor ridge pattern was discovered and found to match the first, I would have been converted.

There were, of course, scientists more conservative than I who would have demanded more evidence that the crust of the earth could move. Some of these suggested that what the instruments were detecting was not the frozen magnetic alignments of the rocks, but some sort of electrical current in the earth's crust. These people would not have come over until the data on the magnetic alignments in deep sea cores was available. But come over they did, and in very short order, too. The *Eltanin 19* data was analyzed in February of 1966; by the meeting of the American Geophysical Union in Washington, D.C., in April of that year the idea of sea floor spreading was beginning to gain supporters. Even Maurice Ewing, director of the prestigious Lamont-Doherty Geophysical Observatory at Columbia University and one of the leading spokesmen for the "fixist" cause, accepted sea floor spreading within three years.

The conclusion to be drawn from this example is obvious: when there is clear and convincing evidence available, the great majority of scientists give up the beliefs of a lifetime and embrace the new idea in an amazingly short time. The key phrase here is "clear and convincing." So long as the picture presented by the data is fuzzy, so long as there are alternatives available, the best that can be expected is that the new idea will not be rejected out of hand. The historical example of continental drift, then, proves exactly the opposite of what the folklore surrounding

Alfred Wegener would lead us to believe. Far from being close-minded, the scientific community accepted his central idea with alacrity, *once there was overwhelming evidence to support it.*

In the light of this conclusion, what are we to make of the longstanding refusal of the majority of scientists to take things like UFOs and psychic phenomena seriously? This is a particularly interesting question to me, because for the past several years I have been involved with a group of scholars who have formed an organization in which serious research on fringe areas in science can be discussed. At meetings of this organization, which includes people from many different disciplines, I have listened to (and argued with) scientists who are trying to do serious work in all sorts of unconventional fields. Over the years, I have become convinced that the general scientific community is absolutely right when it refuses to divert its resources into the study of these areas. Let's take UFOs as an example of what has led me to this conclusion.

Unidentified Flying Objects have been reported in the skies over North America since the latter part of the nineteenth century (at least). This fact in itself isn't too surprising, since there are all sorts of things in the sky that people can't identify at the time of the sighting. Everyone who looks into UFO reports agrees that most of them can be explained in terms of simple phenomena: airplanes, weather balloons, the planet Venus, and so on. The real question is whether there is some subset of the reports that corresponds to objects that are truly unidentified in the sense that they cannot be explained in terms of known physical effects. The most common (but by no means the only) hypothesis is that if such reports exist, they would be evidence for extraterrestrial visitors to earth.

The test facing UFO supporters is very similar to the one that faced Wegener when he first introduced continental drift. Both had to convince their fellow scientists that there were new phenomena to be explained, so that new ideas must of necessity be entertained. We have already seen how Wegener's initial evidence was not equal to this task. The question we must ask, then, is whether the evidence for the reality of UFOs is any more compelling today than Wegener's evidence for continental drift was in 1915. After looking closely at some of the best cases reported by supporters of the reality of UFOs, I must say that

I find their evidence a good deal weaker than Wegener's was. Consequently, I think the great majority of scientists are right when they reject the UFO hypothesis and turn back to their normal research projects.

The basic problem is this: there are thousands of cases of UFO sightings in the files of investigators. Each case, however, "goes soft" when it is examined closely. There is always something fuzzy—always at least one alternate hypothesis that can't be ruled out. Perhaps you can get the flavor of what I mean by looking at one example.

One of the most famous recent UFO cases involved the filming of unusual lights from a plane flying along the coast of New Zealand. It later turned out that the Japanese squid fleet was fishing in the area, and that squid fishing involves the use of very powerful lights to attract the fish to the surface. Some people have analyzed the film and claimed that they have evidence that the lights could not have come from the squid fleet; but after viewing the film and listening to the arguments myself, I see no way that the "squid boat" explanation of the event can be ruled out. Indeed, the only way you can conclude that the lights weren't from squid boats is to make a long string of rather shaky assumptions about the way the film was taken: that the cameraman didn't move the camera for this period of time, that it wasn't tilted a little during that segment, that the boat wasn't swinging around in the water, that there were no surface pockets of fog on the ocean and so on. If any of these questionable assumptions isn't true, the argument fails and you're back to explaining the film in familiar terms.

This is actually a typical analysis of a UFO event. In each case, you wind up asking a question like "Is it easier to believe that the cameraman tilted the camera slightly without noticing it or that there were extraterrestrial spacecraft around New Zealand that day?" Given the kinds of standards scientists apply to these sorts of questions—the kind that they applied in the case of continental drift—the answer is obvious. Very few people are going to invest research time in a field when the evidence is so shaky and the chances of success so small. So, except for a few people who seem to be willing to gamble on a long shot coming through, the field is ignored. And rightly so.

You should note in passing that it is not necessary for scientists to prove that some alternate explanation of a UFO event is true. All they have to do is show that there is some reasonable probabilty that it could be true. As long as this can be done, the criterion of overwhelming evidence that we saw operating in the case of sea floor spreading will not be met and the new hypothesis will be rejected.

But Wegener was proved right in the end, wasn't he? Yes he was, at least in part. There was no way, however, that this could have been anticipated. It is also true that at some date in the future compelling evidence for the reality of UFOs might be found. A fleet of flying saucers could suddenly appear over a major population center, for example, being seen and photographed by hundreds of people in the process. If that happened, supporters would have their equivalent of the *Eltanin 19* profile. But it hasn't happened yet, and to me that implies that it probably won't happen in the future either. Furthermore, until it does happen, scientists are perfectly justified in ignoring the whole area and should not be accused of being close-minded if they do.

After all, as I continually remind my students, a mind can be open without being empty.

# FOUR

# The Earth as a Planet

*In the beginning was the mounting fire*
*That set alight the weathers from a spark*

—DYLAN THOMAS,
*"In the Beginning"*

I T HAS BEEN SAID that one of the most important benefits derived from the space program had nothing to do with technological advances, but was simply the view of our own planet swimming in the emptiness of space. This ability to see our planet from the outside is supposed to have played an important role in the growth of the environmental movement during the 1970s. Whether this is true or not, there is no question that being able to see the earth as an integral part of the solar system has contributed greatly to our understanding of the forces that make it what it is today.

It's always difficult to come to reliable conclusions when you have only one thing to study. We can learn about the earth's weather patterns, its climates, and its plate tectonics, but we can never be sure we understand the basic principles behind any of these things until we test our knowledge on something other than the earth. To understand the earth, in other words, we must leave it. There are many examples of how this process has worked. Computer codes developed to model the sandstorms that sweep across the surface of Mars are now routinely used to predict long-range climate effects on earth, and are playing a major role in the ongoing debate about whether a nuclear winter would follow a major nuclear war. Because of this new insight, we also have a much better appreciation of the earth's unique role as the sole abode of life in our solar system (and perhaps in the entire galaxy).

To get this sort of perspective on our own planet, we have to go back to the very beginning of the solar system, when the sun and planets were forming out of a contracting cloud of interstellar gas. As each bit of material in the cloud was pulled toward the others by the force of gravity, the cloud collapsed and heated up. Any small rotational motion the cloud might have had at the beginning would be magnified during the collapse, for much the same reasons that an ice skater spins around much faster when he pulls in his arms. At the center of the cloud, the temperatures and pressures rose to the point where atomic nuclei were forced together and a nuclear fusion reaction was ignited. Eventually the energy streaming outward from the central core provided a force capable of counterbalancing gravity, and the contraction stabilized. A star—our sun—was born. At first it sputtered and stalled, like a balky car engine on a cold morning, but soon began to pour a steady stream of energy out into space.

With regard to the amount of matter or energy involved, the formation of the sun was the most important event involved in the collapse of the cloud. While the sun was forming, however, some of the debris from the process of construction spread out in a thin disk around the central mass. The centrifugal forces associated with the rotation of the disk balanced the inward pull of the sun's gravity, so that the debris continued to circle in orbit. At this point the solar system looked something like a gigantic version of the planet Saturn: a central mass surrounded by a small amount of material spread out in a thin disk. It was from this debris that the planets eventually formed.

During the contraction, the general heating of the material toward the center of the cloud led to a separation of material within the ring. In the inner reaches of the disk, the temperature was too high for materials like water and methane to form into solid chunks. On the other hand, materials with a high melting point, such as iron and silicon, could remain solid everywhere. As the sun was forming, grains of solid matter in the disk were colliding, sticking together, and forming small objects up to a few miles across. Some of these small bits of material survive today in the form of asteroids and meteorites. They are referred to as planetesimals, whether they have survived to the present or not.

The present theory is that the planetesimals collided with one another, growing into larger masses. The largest of these growing bodies began to "eat up" their neighbors, picking up material as they moved through the swarm of planetesimals the way a windshield picks up insects on a summer night. The ignition of the nuclear fires in the sun and the subsequent sputtering start-up occurred sometime while this consolidation was going on, and the outpouring radiation removed the gases and small bits of material from the inner parts of the solar system.

As far as we can tell, the actual spot where a planet formed was simply a matter of chance; wherever a large body happened to assemble itself through the influence of gravity, it continued to grow. The formation of the planets was a classic case of the rich getting richer, although in this case the poor didn't get poorer, they just got gobbled up. The formation of the earth 98 million miles from the sun, therefore, was not a necessary consequence of any law of nature. It just happened that a series of more or less random events caused the cloud of material 98 million miles from the sun to be a little more dense than its surroundings. Given this fact, the eventual formation of a planet at that spot was inevitable. The earth, however, could just as well have formed 90 or 100 million miles from the sun.

This picture of the formation of the planets explains why the inner planets are small and rocky while the outer planets are large and gaseous. The inner planets formed in a region in which only materials like iron and silicon could remain solid, while in the cooler outer regions occupied by the gas giants, materials like methane and ammonia were still available as building blocks. The rocky members of the solar system are generally referred to as the terrestrial planets, while the gas giants are called the Jovian planets (after Jupiter, the largest of them). Because the earth's moon is so large and so available for study, it is usually included in the list of terrestrial "planets," although it is, strictly speaking, a satellite.

Now let's shift our point of view from that of an observer outside the solar system, seeing everything happening at once, to a point on the surface of one of the large bodies that would eventually become the earth. From that vantage point, we would see the newborn sun glowing off in the distance, somewhat less bright than it is today. The most important aspect of our environment would be the constant rain of material from the sky. As the earth accumulated matter by sweeping up material from its orbit, we would see a rain of meteorites unlike anything we have encountered since. The fall would be so rapid, and the impacts so devastating, that the earth would begin to warm up. This warming trend would be augmented by the decay of radioactive nuclei throughout the body of the earth—a process that is still going on today. The earth would continue to accrete debris from its orbital neighborhood until it had added to itself all that was available. Like radioactivity, the process of accretion is still going on today: every time you see a shooting star, you are seeing a piece of material the size of a grain of sand being added to the earth.

Once the meteorite bombardment and radioactivity had heated the earth and other terrestrial planets, some simple laws of nature can explain what happened next. It is a well-known fact of everyday life that once an object is heated, it takes a while to cool off, and that during the cooling-off period it transmits heat to its surroundings. This is the principle behind homey devices like the old-fashioned bedwarmer, in which hot coals placed inside a pan were set under the blankets in unheated bedrooms. The coals cooled off slowly by giving off heat to their surroundings (in this case the bed and covers themselves). The net result was that when someone climbed into the bed, he or she avoided those horrible few minutes that would have passed while body heat brought the bed to a bearable temperature.

The same general principle is used in modern active solar-heated houses, as well as in old adobe and brick buildings in the American southwest—buildings that used to be just "old-fashioned" but are now graced with the title of "passive solar structures." In both the new and old systems, heat from the sun is stored during the course of the day and released into the house during the night. In modern systems, the heat is gathered in collectors on the roof and pumped to a storage system (usually in the basement). The storage system is normally a large tank of water or a bin filled with rocks. During the day the storage material heats up as warm air or water from the roof runs through it, and at night it gives its heat up to the air or water that runs the heating system in the house.

Passive solar buildings actually work the same way, although they incorporate the heat storage reservoir into the structure of the building. Sunlight streaming through a glass window wall during the day can heat a thick slab of tile or cement on a floor; then during the night the heat is automatically released into the house.

All three of these examples—bedwarmer, modern active solar heating, and passive solar buildings—operate according to the same laws that governed the behavior of the planets after the meteor impacts had started. All the planets were subjected to the same process of bombardment as the earth. But, as we shall see in a moment, the way they cooled had a profound influence on the forms they would have throughout their lifetimes.

Any bit of material that is heated will give off heat to its surroundings. Consider, then, two separate pieces of heated material in a new planet (see figure 4–1). One piece is in the interior of the planet, surrounded by similar material; the other is on the surface, with open space or atmosphere on one side. Both pieces will lose heat to their surroundings. The piece in the interior will transfer heat to its neighbors, but—and this is

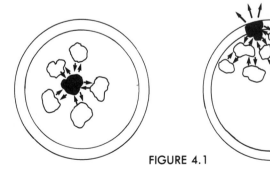

FIGURE 4.1

the crucial point—it will also have heat transferred back to it from its neighbors. Therefore, its net heat loss (and net drop in temperature) will be small or even zero. The piece on the surface, on the other hand, will radiate heat to the outside without receiving any heat in return. Consequently, the piece on the surface loses heat to the outside and will cool off. Eventually, heat from the interior pieces will be transferred to the surface and radiated away—this is, in fact, a detailed picture of how a large body cools. The important point is that while heat can be stored anywhere in the body, it can only be lost through the surface.

Your experience with a campfire can quickly convince you of the truth of this statement. If you want to make the campfire last until morning, you bank the coals, piling all the glowing embers into one heap at the center of the fire. This has the effect of making many of the coals into interior pieces, like the one in our example. These coals will transfer heat among themselves, but will stay hot for a long time. They will lose heat only through the relatively small number of surface coals.

If, on the other hand, you want to put the fire out, you take a stick and spread the coals over the ground. This has the effect of eliminating interior coals completely—every coal is now on the surface. Consequently, they cool off quickly and the fire ceases to be a danger to the forest around it.

In the language of the physicist, all of these arguments can be summed up in the statement that the rate of heat loss in any object is governed by the surface-to-volume ratio. The larger the surface of any object relative to the amount of material it holds, the more quickly it will lose heat and cool off. In the case of spherical bodies like planets, this means that the larger the radius of the planet, the slower will be the rate at which it can radiate its heat into space.

Take the earth and Mercury as an example. The earth has a radius of 4000 miles, Mercury about 1400. This means that the earth has a volume roughly eighteen times that of Mercury, but only seven times the surface area. Since heat can be accumulated throughout the volume but is lost only through the surface, the earth will have a much harder time than Mercury in shedding heat that is added to it. When the meteors fell and the nuclei decayed early in the formation of the solar system,

Mercury could radiate the accumulated heat out into space much more easily than the earth, simply because Mercury had so much more of its material on the surface.

We haven't taken any measurements of the interior of Mercury yet, of course, but we do have rather extensive seismic data on a terrestrial planet of about the same size—our own moon. Apollo astronauts left behind a number of remote seismic instruments on the lunar surface, and by analyzing the signals that accompanied small earthquakes (moonquakes?), geologists were able to show pretty conclusively that the moon is made up of materials that had melted down to a depth of just over 100 miles; but below that it was more or less in the state in which it was originally assembled. What this means is that the moon was able to shed the heat associated with the meteorites and radioactive decay almost as fast as it was brought in. Enough heat accumulated to raise the temperature of a thin layer on the surface above the melting point, but that's all. The heat was sent out into space before it could raise the temperature of the body of the planet. Within a few hundred million years, the surfaces solidified again, wrapping the planet in a single, continuous lithospheric plate.

On the moon, we know that during the molten period lighter materials floated to the surface, so that when the newly solid crust was shattered by some of the remaining large meteorites, molten lava flowed into the resulting craters to form the "mares" (Latin for seas) that can be seen on the moon with the naked eye. The most recent of these, Mare Orientalis, was formed some 3.9 billion years ago—just a few hundred million years after the formation of the earth-moon system. Mercury and the moon, then, represent planets in which the surface-to-volume ratio is high. They are analogous to a fire in which the coals have been spread out pretty thoroughly, and which therefore cannot generate a lot of internal heat. In this situation, there is no tectonic activity, jut a solid, unbroken lithospheric plate surrounding the entire body (although, to be complete, we should note that Mercury's plate appears to have been fractured in a few spots, perhaps by tidal interactions with the sun).

The other terrestrial planets—Venus, Mars, and the earth—are larger than the moon and Mercury. They are more analogous to fires in which the coals have been well banked. If heat

63

is introduced into such a system quickly enough, it simply doesn't have sufficient surface to escape through. Consequently, it diffuses through the interior, raising temperatures throughout the system. In the case of the earth, most geologists believe that enough energy was added during its first few hundred million years to melt the planet clear through. During this molten period, the earth underwent the process known as differentiation. Heavy minerals (like those containing iron) sank toward the center under the influence of gravity. Lighter materials (containing elements like silicon and magnesium) floated to the top. During this stage the earth was like a bottle of oil and water that has been thoroughly mixed and then allowed to sit quietly. The oil separates and comes to the surface, while the heavier water sinks. (You can experiment with cooking oil and water to see this for yourself.) In just the same way the earth's heavier materials sank and the lighter materials rose, giving the earth an iron core.

The fact that the earth has become differentiated explains why mining is such a difficult business. Most of the heavy minerals like iron and gold long ago sank deep into the earth; only a few odd remnants are left near the surface where they are accessible for exploitation. In fact, the richest accessible mineral deposits in the solar system are not on the surface of any planet or moon, but in the asteroid belt between Mars and Jupiter, where large rocks of undifferentiated, primordial material circle the sun. Perhaps someday we'll be able to exploit those deposits and leave our earth alone.

Mars and Venus are large enough, and have a small enough surface-to-volume ratio, to have melted and undergone differentiation similar to that which we've described for the earth. Mars, being the smallest of the three, was able to dump its excess heat most quickly. Like the moon and Mercury, it quickly formed a single global lithospheric plate, and its subsequent geological activity appears to have been mainly volcanic in nature. Venus is only a little smaller than the earth. NASA radar maps taken through the Venusian clouds show a definite two-tiered structure, with extensive highland interspersed with what would be ocean bottoms if the surface temperatures weren't well above the boiling point of water. There might even be a set of fold mountains on the surface. As far as we can tell, however, there is no tectonic activity on Venus today, although there are probably active volcanoes.

To summarize, the smaller terrestrial planets have a high enough surface-to-volume ratio so that the initial inflow of heat from meteorites and radioactive decay was sufficient to melt only a layer at the surface. The rest of the terrestrial planets, being larger, were melted completely through by the same set of events.

Radioactive decay continued to pump heat into the interiors of all of the terrestrial planets even after the initial meteor showers had stopped. Mercury and the moon, as you might expect, simply allowed this heat to flow out into space through their surfaces. Mars, although it has too low a surface-to-volume ratio to have disposed the heat from both radioactivity and meteors, nevertheless was capable of handling radiation alone. Consequently, for the last few billion years Mars has been able to dump its interior heat into space like the spread-out coals in our campfire. During this period, the planet has been slowly cooling off. Its surfaces are now covered with a single lithospheric plate.

Venus is something of a special case. It is significantly larger than Mars, and only slightly smaller than the earth. As far as we can tell, the surface of the planet consists of a single plate. Like Mars, Venus appears to have no tectonic activity. On the other hand, there are probably active volcanoes on Venus, and some planetary scientists speculate that there may be plumes of hot magma coming up to the surface at various points on the planet. Thus, Venus appears to be in the late "cooling-off" stage of development; it has no tectonic activity now, but there are still heat-driven geological processes at work.

The earth is the largest of the terrestrial planets, and consequently has the largest surface-to-volume ratio. In terms of our analogy, the earth is like a well-banked fire, with most of its coals in the interior. Thus, after the initial molten period, the earth parted company with Mars and Venus. The latter were able to deal with their interior heat by the simple process of diffusion, an option that wasn't available to the earth. Its heat had to be removed from the interior by some more efficient mechanism.

That's where convection and plate tectonics comes in. You

can picture the planets as being analogous to a pot of water on the stove. When the stove is on a low setting, the heat can flow up through the water and radiate out through the surface without any motion in the water. You can feel the heat with your hand if you put it near the water. If you turn the stove up, however, this relatively quiet method of heat transfer won't suffice; the heat simply overloads the system. The only way the water can get rid of the accumulating heat is to start to boil. As we saw in chapter 1, convective processes like boiling provide a rapid transfer of heat through any material.

The other terrestrial planets are like the pan of water when there is no boiling. Their internal heat diffuses out through the surface, perhaps with a little help from volcanoes, but there is no need for the material to move. The earth, on the other hand, is like the pan of water with the setting turned high. In order to get rid of the heat being introduced continuously beneath the surface, some sort of convective action has to be set up. Upwelling material from the interior carries heat to the surface, cools, and sinks back to repeat the cycle. This continual churning prevents the formation of a single plate covering the entire surface, so that the surface of the earth is in a constant state of turmoil. Alone among its sister planets, the earth has a surface that has not frozen for all time.

As far as we can tell, this unique feature of the earth's structure is, in one sense of the term, accidental. The earth's radius is only about 5 percent larger than that of Venus, and the earth is only about 15 percent more massive than its sister planet. Yet this small difference in size, probably caused by the fact that the initial disk of debris from which the planets formed was slightly more dense in the region where the earth formed, led to an immense difference in later geological behavior.

A number of consequences follow from the fact that the earth alone among the planets has continuing tectonic activity. For one thing, it means that our earth is still evolving, while the other planets have more or less settled down into their final state. It would be hard to believe that the continuous change of environment associated with the movement of tectonic plates had no effect on the development of life on earth. On the contrary, it is reasonable to suppose that the constant change in environment provided by tectonic motion was an important driving force in the development of intelligence. Many evolutionary biologists have argued, in fact, that constant challenges and environmental changes are necessary for such developments. Given this state of affairs, the accidental origins of the earth's tectonic properties have important philosophical implications. It's not too far-fetched to suppose that only tectonic activity could create an environment that provided challenges on a time scale sufficiently long for living systems to be able to adapt, yet also provided them often enough to force intelligence to develop. If this is the case, then the earth might be unique not only because of its tectonic plates, but because it might be the only planet in the galaxy with the right conditions for developing intelligence, conditions that depended in part on chance. It's an interesting thought.*

Another consequence of the earth's place as the only planet with tectonic activity is that our planet is the only one with true mountain chains. The Rockies and the Appalachians are more than interesting—they are unique in the solar system. Other planets have large volcanoes, and the moon and Mercury have jagged edges on their large craters; but the idea of a mountain as a living, changing, evolving structure is unique to the earth.

Someday, of course, the earth will lose this distinction. The end product of radioactive decay is always a series of stable, nonradioactive nuclei. If the earth survives long enough, most of the radioactivity in its interior will spend itself and the heat input will die out. In terms of our analogy, the setting on the stove will be turned down and the water will stop boiling. Heat will flow through the solid earth as it does on Venus and Mars and the outer regions of the earth will start to cool. The continents will be frozen into whatever arrangement they happen to have at that time, and the tectonic plates will be welded together into a single plate that circles the globe. The world will then be as it has been falsely imagined to be throughout so much of human history—static and unchanging.

I've never seen a calculation about when to expect the end of tectonic activity on the earth, but from the fact that the most abundant radioactive element—uranium—has a half-life of over

*For a more complete discussion of the unique properties of the earth and the implications for the development of intelligence, see *Are We Alone?* by Robert Rood and the author (Scribners, 1981).

four billion years, I expect that we'll see the sun go out long before we see the plates stop moving.

## A Word About Meteorites

Looking at the earth as just one special case of planetary evolution is interesting, but you may be wondering at this point how it is that we can know so much about events that happened many billions of years ago. There are two sorts of answers to this question—one philosophical, one practical.

At the philosophical level, we can point to one of the most important discoveries of modern astronomy. If you look at a galaxy billions of light-years away, you are seeing that galaxy as it was when the light left it, billions of years ago. Looking out, in other words, corresponds to looking backward in time. Our studies of distant galaxies show that the same laws of physics that operate in our laboratories today governed the behavior of objects in the distant past as well. The laws of physics seem to be valid not only throughout all space, but throughout all time.

This means that when we examine the coals of a campfire to understand the way that heat moves in and out of material bodies, we are not only discovering things about the behavior of stars in distant parts of the universe, but we are discovering things about the past history of our own planet as well. This is an astounding fact—one that deserves to be shouted from the rooftops. It is only long familiarity, I suspect, that makes us so blasé about it. In terms of understanding the formation of the earth, this universality of the laws of nature tells us that we can study the behavior of heated bodies in our laboratories, and by extrapolating the knowledge thus gained, we can deduce the origins of our own planet.

At the practical level, we can learn a good deal about the formation of the solar system because some of the leftover construction materials are still around in the form of small chunks of material orbiting the sun. Occasionally, one of these bits gets too close to the earth, finds itself trapped in our gravitational field, and comes plummeting into the atmosphere. Most of the time the falling bodies burn up during this descent, leading to the familiar phenomenon of the "shooting star." Sometimes, however, there is too much material to burn up and some of it actually reaches the surface of the earth, in which case it is called a meteorite.

Think for a moment about what this means. The blackened piece of rock you see in a museum was preserved in its present form for more than four and a half billion years. While its brothers were incorporated into planets, melted, and differentiated, this particular chunk of rock patiently orbited the sun, absorbing the occasional cosmic ray, until it fell to earth. It is a messenger from our past, carrying priceless information about what the solar system was like before the sun ignited and the planets formed.

In 1969, for example, an asteroid about the size of a car entered the earth's atmosphere, broke up, and littered the area around the town of Pueblito de Allende in northern Mexico with fragments. The "Allende" meteorite, as it is now called, has given us our best information about the process which started the collapse of the dust cloud that eventually became the solar system. Examining the materials in the meteorite, scientists discovered that they contained a slightly higher content of a certain isotope of magnesium than one would ordinarily expect to find. The particular species of magnesium involved arises from the radioactive decay of a type of aluminum produced copiously in a supernova. (We'll discuss supernovae in more detail later, but for the moment you can think of a supernova as an explosion accompanying the death of a large star.)

The picture that emerged from the study of the Allende meteorite was stunning in its simplicity: A little over four and a half billion years ago a nearby star died, blowing its inventory of material out into the galaxy. Some of this material included atoms of aluminum that would someday decay into the particular isotope of magnesium mentioned above. The shock wave resulting from the explosion started the cloud's collapse, and the material from the star mingled with the material already in the cloud. The alien atoms were incorporated into the sun and into minerals in the disk of debris from which the planets formed. Most of these minerals were incorporated into planets, where they were diluted beyond the point of detection. In a few cases, however, they were part of bodies that escaped the planetary formation process and became asteroids. As the eons passed, the original atoms of aluminum decayed and became magnesium. Finally, quite by chance, one of these asteroids fell to earth, bringing with it hints of the story of the beginning of the solar system.

Other meteorites have yielded up organic molecules, revising our ideas about how chemical reactions occur in space. Still others bear evidence that they were once on the surface of the moon or Mars, partly melted, then kicked back into space by meteor impact. Each visitor from space has its own tale to tell, and each is potentially unique in the knowledge it can bring us.

This is the reason why it is extremely important that no meteor be lost to science because someone picks it up as a souvenir. Even universities and state agencies have been known to lose or misplace meteorites entrusted to their care. If you should happen to have the good fortune to come into possession of a meteorite, I strongly urge you to write to

The Curator of Meteorites,
Smithsonian Institution,
Washington, D.C. 19560

He will see that it is properly dealt with.

# FIVE

# Where Do All Those Rocks Come From?

*There was rock to the left and rock to the right,*
*and low lean thorn between*
*And thrice he heard a breech bolt snick, tho*
*never a man was seen*

—RUDYARD KIPLING,
*"Ballad of East and West"*

I WAS A LITTLE SURPRISED when I started to slide. I had been working my way across what geologists call a talus formation: a sloping pile of debris that results from the breakdown of material higher up on a mountain. At first, the slope had consisted of large rocks; but as I moved across, the material underfoot got smaller and smaller. The other members of the party, being lighter and perhaps wiser in their choice of path, encountered no difficulty, but I eventually reached the point where even extreme measures like lying flat on the slope and digging my feet in did no good. Backpack and all, I started a slow, inexorable slide down toward the base of the mountain, 200 feet below.

I wasn't in danger—I had on sturdy clothing and was sliding too slowly to suffer any harm. The damage was likely to be only to my dignity. Nonetheless, as I felt the gravel give way and looked down to see where I was headed, the thought "What the hell am I doing up here on these rocks?" did flash across my mind.

Actually, the situation wasn't bad at all. I was able to aim my downward movement toward some shrubs, grab them as I went by, and eventually scramble up and rejoin the group. Later, after a hiker's lunch of apples, sausage, and cheese (not to mention a few beers), I was ready to take a more philosophical view of the situation.

I had come to this particular spot to look for fossils. It was the only place within a hundred miles where they were to be found, and they were here in profusion—seashells, castings, even some fossilized fish fins. The local scene was very different from the landscape we'd driven through to get there. Instead of the massive, angular, dark rocks that made up the surrounding mountains—rocks that had barely been weathered during the millions of years since the mountains had formed—the rocks here were soft, pitted, and crumbly with the effects of weather. They were laid out in sheets: looking at exposed areas was like looking end on at the pages of a book. The rock was clearly layered, and very different from anything else in the neighborhood. I realized that I had asked the wrong question during that moment of crisis on the slope. The interesting question was not what I was doing on the rocks, but what were the rocks doing under me?

You don't have to travel to remote areas of the Rockies to discover how different one kind of rock can be from another. Every place where rocks—the "bones of the earth"—come to the surface, they are likely to be different from their neighbors. Sometimes, where a road cuts through a hill or a construction crew has emptied out a deep hole for the foundation of a new building, you can see the sort of layered structure I saw that day in the mountains. A few miles away, in the same situation, you may see a hard, impenetrable gray stone with no discernible layers at all. Any handful of individual rocks you pick up off the ground shows a wide variety of types. Some are sandy-colored and rough to the touch, others hard, smooth, and glassy. Some are solid colors, others show streaks and speckles. Some are dull, others very beautiful. Where do they all come from, and why do they differ so much among themselves?

The appearance of a rock, like the appearance of a person,

tells you something about the sort of life it has had. Geologists put rocks into three general categories: sedimentary, igneous, and metamorphic. The names refer both to the way the rock was formed in the beginning and the things that happen to it afterward. Sedimentary rocks begin to take shape when material settles out of water, forming layers on the bottom of a lake or ocean. They have the characteristic appearance of stacked sheets mentioned above. Igneous rocks are formed by the cooling of melted material. Once formed, both igneous and sedimentary rocks will retain their original composition and appearance provided nothing unusual happens to them. But if they are subjected to heat or pressure or other disruptive agents, they can change form. In that case, they are classed as metamorphic rocks—rocks that have undergone metamorphosis.

One of the great geological debates of the eighteenth century took place over the question of the origins of the earth's rocks. One side—the "Neptunists"—held that all rocks were formed by the process of sedimentation in the oceans. The other side—the "Vulcanists"—held that all rocks were formed by processes like the cooling of lava. Neither side had deduced the existence of metamorphosis, and both adopted what we would call an extremist position. The Neptunists, for example, pointed to the existence of fossils in some rocks, arguing that nothing could have lived in molten lava. The Vulcanists pointed to the indisputable fact that volcanoes like Vesuvius and Etna in Italy had, within recorded history, produced lava flows that had buried entire cities under newly formed rock.

I suppose if you had had to pick a winning side in this debate, you would have had to say that the Vulcanists were closer to the truth than their opponents, but not by much. If you think back to our discussion of the formation of the earth, you will recall that very early in its history the entire earth went through a stage when it was molten. Thus, in the very beginning, before any geological processes had had a chance to start, every solid object on the surface formed from cooling and solidification of heated material, much as rocks today form when volcanic lava cools off.

But once the earth's crust solidified and the oceans formed, the steady process of weathering began. Even without the presence of plants, the action of rain and changes in temperature would start to knock little chips and grains off the rocks at the surface. The surf at the edges of the newly formed oceans would do the same. Relentlessly, the rocks would be attacked and broken down. Water sluicing down would carry this debris to the oceans. There the water would slow down and stop, and the load of rock chips, dust, and grains would gradually settle to the bottom. As time went on and more sediment piled up, the original grains would be packed together more and more tightly. In some places, chemical processes in the ocean caused small amounts of crystalline material to precipitate into the spaces between the grains and form a kind of cement that held them together. The effect was roughly the same as you'd get if you poured a bottle of glue over a pile of sand or dirt. The combined effects of the pressure and the chemical action turned the loose assortment of sediment into a rock. As you might expect, the rock would be classified as sedimentary. Sandstone is a common type of sedimentary rock formed from sand grains, while shale is another type formed from finer particles of clay. Later on in the earth's history, when life had developed in the oceans, sediments were formed by a downpouring to the bottom of the skeletons of tiny marine organisms. Limestone is a common rock formed from this sort of sediment: its white color derives from the bones of the billions of organisms whose remains make up the rock.

The key act in the formation of sedimentary rocks, then, is the settling of material to the bottom of a body of water. You don't have to be a professional geologist to see this process in action. Next time you get a heavy rain, watch the runoff patterns. You'll see the rain running rapidly down sloping surfaces, but often the water will back up to form a large puddle where the ground levels off. The most spectacular example of this effect occurs when the early rain washes a lot of debris into a drain somewhere, forming a porous dam and forcing the water to back up behind it. When this happens, sediment being carried along in the rapidly running water will start to settle in the stationary water of the puddle. The next day, when the water has had a chance to run off, you'll see a layer of dirt where the puddle used to be. This dirt is the first layer in the sedimentation process. If the settling went on for millions of years instead of a day or so, that dirt would eventually be incorporated into a shale-like rock.

Because sedimentary rocks are formed in bodies of water, the shape of the rocks will be that of level beds. A thick bed of sedimentary rock will have that characteristic layered appearance. In fact, the best example of a sedimentary "rock" I've ever seen was what collected in the bottom of a wheelbarrow when I was building my house. I had used the wheelbarrow to mix various mortars and cements and, being occupied by other chores, never bothered to clean it out. Eventually, I had to turn the whole thing over and beat it with a hammer. What I found when I finished dislodging the encrusted cement was a "rock" composed of different-colored layers, each representing a different task—the deck foundation, the sidewalk, the hearth, etc. When I broke it with my hammer, it looked just like the formation you see when a deep layer of sediment is cut away by a stream—a miniature Grand Canyon. The only difference was that my "rock" built itself up over a span of months, while the Grand Canyon took millions of years.

The fact that sedimentary rocks always form by settling is important because it tells us that regardless of the rock's present configuration, it must have been formed in horizontal layers. Thus, when we see situations like those shown in the photographs on pages 8 and 9, we know that some powerful forces must have been at work *after* the rock was formed. In fact, it was the existence of tilted and deformed sedimentary rocks that helped to convince geologists in the nineteenth century that there were forces in the earth capable of lifting mountains. In the examples shown in the photographs, the forces involved were those that created the Rocky Mountains and the Black Hills, respectively.

So although the material in sedimentary rocks ultimately came from rocks forming out of the cooling of the newborn earth, it is clear that the main characteristics of sedimentary rocks are related to the time they spend on the bottom of oceans or lakes. Therefore, the Neptunists were close to the truth when they argued that some rocks came from the sea. At the same time, it is also true that not all rocks are sedimentary. We know that some of the time, molten material from the earth's interior—material called magma—comes to the surface and cools, forming a totally different kind of rock. Indeed, it is this process that drives tectonic change and forms the ocean floor.

There are two ways in which igneous rocks—rocks formed

from heat—can be created. The most spectacular, of course, is in volcanoes, where lava flows create blankets of new rock. Less dramatic but equally important are occasions when hot magma moves up from the depths through faults and cracks in the earth's crust, though never reaching the top. The magma cools to create underground formations. In the course of weathering, these formations can become exposed and form part of the visible landscape. The Sierra Nevada Mountains in California are an example of this process, but igneous intrusions need not be grand and massive. Geologists regularly catalogue igneous formations that are only a few inches across.

Once sedimentary and igneous rocks have formed, they are subject to the same forces of uplift and tectonic motion that we discussed earlier. The fact that a sedimentary rock is formed on the ocean bottom does not mean that it will be found at the same low altitude at some later time. The rocks that precipitated my downward slide on that slope were sedimentary, but they were on a plateau near Yellowstone National Park, some 8000 feet above sea level. The uplift that formed the Rockies had raised them from the sea floor to the mountaintop. This example illustrates a common pattern: one set of processes governs the formation of a rock; another acts in other ways to govern the rock's life history.

This lesson is particularly important when we consider the third great class of rock types—the metamorphic. Once a rock is formed, it lives out its life in an environment that affects its form. For example, when we discussed the first step in sedimentation, we assumed that a piece of rock exposed to the elements would be gradually broken down. Instead of remaining as a large, solid block of material, it would be decomposed into individual grains, and these grains would eventually be incorporated into ocean floor sediments. This is the simplest example of how a rock's environment affects its form. There are, however, many factors other than weathering that can affect a rock. If these changes occur within the earth, and if they don't involve melting, we say that the rock undergoes metamorphic change. Metamorphic transformations takes place in various ways.

If the rock is heated, the atoms in its crystal structure are pushed apart and may actually break loose from their normal chemical bonds to rearrange themselves into a new mineral type. If the rock is subjected to pressure, the effect is the opposite— atoms are pushed closer together. This often happens at the bottom of very thick layers of sediment. The atomic structure of a rock can also be changed by the migration of different species of atoms into the rock, where they replace some of the old atoms and, in the process, alter the atomic structure. Some-times, all three agents—heat, pressure, and chemicals—can act together. Large areas of the earth's surface have undergone such changes. Much of the rock in the Alps and parts of the Northern Rockies is metamorphic.

Rocks will undergo metamorphosis only so long as one of the agents of change continues to act. When the agent is re-moved, the change stops and the rock stays in its new form until subjected to another agent. This is for some people a difficult notion to grasp; their first impulse is to suppose that as soon as the pressure or heat is removed, the rock will revert to its old state, like a rubber band snapping back when you let it go. What one must remember is that metamorphosis actually alters the atomic structure of the rock, and that atoms will not rear-range themselves unless there is a force driving them to do so. Metamorphosis is in fact much more like what happens when you stretch a rubber band beyond its elastic limit. It either breaks or loses its elasticity, the atoms in the rubber having been permanently disturbed. The band will not pull itself back into its original shape when you let go. It has lost its snap and will never get it back again. In the same way, marble (a metamorphic rock) will never revert back to its original limestone, no matter how long you wait.

So, the question "Where do all those rocks come from?" has a complex answer. Rocks may be formed in the melting or sedimentation process, but they can also change their form slowly in response to their environment. Thus rocks are not the static, lifeless things they are often imagined to be, but are just like living things—constantly changing with time. What we see in a particular location at a particular moment depends not only on how the rocks originated but also on what has happened to them over the last few hundred million years.

So where do the sedimentary rocks that prompted my ques-tion in the first place come from? Their history includes their original formation at the bottom of the ocean that once covered the western states, being uplifted when the Rocky Mountains were formed, and then having nearby sedimentary rocks scraped off by repeated passages of glaciers. The end product was a butte that slowly weathered, with the bits and pieces of rock accumulating at the base to form the slope I tried to cross.

That certainly answers my question at one level. There are, however, other levels that bear investigation. For example, when a rock melts or solidifies or undergoes metamorphosis, what is happening is that the atoms that compose the rock are being rearranged to make a new mineral. All of the many ways that rocks are created and changed during their life can be thought of as nothing but rearrangements of atoms found at the earth's surface. So when we ask "Where do all the rocks come from?", it isn't enough to talk about sedimentary, igneous, and metamorphic origins. We must go deeper and ask about the origin of the atoms that make up those rocks, and this leads us into some very interesting areas indeed.

Like the rocks and minerals that contain them, atoms have a complicated internal structure. Most of the mass of the atom is located in a dense, positively charged nucleus, while negatively charged electrons circle in orbits, much as planets circle the sun. Usually, there are as many negatively charged electrons as there are positive charges on the nucleus, so that the atom, considered as a whole, is electrically neutral. When this is not the case—if, for example, the atom has lost an electron or two—there is a net positive charge and we refer to the system as an ion.

The atom's nucleus itself has an intricate structure. For our purposes, we need only note that it is constructed from two types of particles: protons (which carry a positive electrical charge) and neutrons (which carry no electrical charge). To remain electrically neutral the charge on the electrons in orbit must exactly cancel the positive charge of the nucleus; and since the electrical charges on the proton and electron are equal in magnitude though opposite in sign, it follows that in normal atoms, the number of electrons will be equal to the number of protons in the nucleus.

Now, it is a fact of subatomic life that there are always loose electrons wandering around somewhere in the world. If a positively charged ion is around long enough, it will eventually encounter one of these loose electrons, pick it up, and be transformed into a run-of-the-mill neutral atom. Thus, any nucleus, once it is formed, will automatically acquire its quota of electrons—a quota which (as we saw) is determined by the number of protons in the nucleus. When we talk about the origin of the atoms, then, what we are really talking about is the origin of the nuclei. The addition of the electrons, though important, is something like an afterthought.

If we start with a collection of protons and neutrons, the process of bringing them together to build a nucleus is complicated by two important facts. First, neutrons are unstable. They will spontaneously change into a proton, an electron, and another kind of uncharged particle in about ten minutes unless they are inside a nucleus. This means that there is a time limit in nature: once a free neutron is created, it has only a few minutes to get to the shelter of a nucleus before it comes apart. We can't keep a neutron around for a thousand years waiting for a nucleus to come along and pick it up.

The second complication has to do with what happens when we try to bring two protons together. Since both protons have a positive charge, the laws of electricity tell us that they will repel each other. If you had to sit two protons down side by side, the repulsive force would cause them to fly apart. The only way to overcome the repulsion is to have the two protons move toward each other at high speed. In this situation the repulsive force may slow them down a little, but it can't stop them from coming together. (This repulsive force does not, of course, exist between protons and the uncharged neutrons.)

So, starting with a collection of protons and neutrons, the first thing we could do to build nuclei is to let the two species of particles combine in pairs. The result would be a group of nuclei consisting of a single proton combined with a single neutron. This "baby nucleus" (as nuclear physicists call it) is called the deuteron: it is the nucleus of the deuterium atom. Since it contains only one proton, it will acquire only one electron. This means that it will have the same number of electrons as an ordinary hydrogen atom (whose nucleus is a single proton), but will be about twice as massive, because the neutron has approximately the same mass as a proton. Atoms like hydrogen and deuterium, which have the same numer of protons (and electrons) but different numbers of neutrons, are said to be isotopes of each other.

Once we've made deuterium, we can go on to try to combine the newly formed systems into more complex nuclei. Since our building blocks now all contain one proton, and therefore have a positive electrical charge, we are going to have to deal with the repulsive force we discussed above. As we saw, this force can be overcome if our deuterons move very quickly. So it comes down to this: to go beyond the simplest possible nuclei, the particles must be in a region where they are moving fast enough to overcome the electrical repulsion between protons. Since there is a direct correspondence between the speed at which particles move and the temperature of the medium composed of those particles, this leads us to the conclusion that the only place we are likely to see complex nuclei made is in regions of very high temperature.

There are, in fact, places in nature where temperatures are high enough for complex nuclei to be forged out of simpler ones. This process goes on in the center of stars like the sun, where the combination of protons and neutrons into alpha particles (two protons and two neutrons) supplies the energy that makes the star shine. In a sense, then, it is the building of nuclei that supplies the energy for all life on earth.

There are two other high-temperature regions where elements have or can be made, although they maintain those temperatures for relatively short periods of time. One short period of element building took place during the first few minutes of the Big Bang, before the universe had cooled to the point where protons could no longer be forced together. The other occurs occasionally when a very massive star "dies" in a catastrophic (and poorly understood) event called a supernova. All atoms with nuclei more complex than hydrogen—which is to say the atoms in all familiar materials—were made in one of these three situations.

Back in the 1940s, when scientists first started thinking about the problem of where large nuclei came from, it was generally believed that all the elements were made at the beginning of the universe, during the hot stages of the Big Bang. The first attempts to build a Big Bang cosmology were in fact directed at explaining the existence of heavy elements. The idea was that a certain quota of complex nuclei were manufactured early on, and that the universe has been more or less existing on its nuclear capital every since.

Unfortunately, this idea didn't work out. The Big Bang was hot enough to make deuterium and alpha particles, but it cooled off too fast for more complex nuclei to be formed from these simple ones. The basic problem was this: processes like the one we used as an example above (protons and neutrons to make deuterium, two deuterium nuclei to make alpha particles) could produce nuclei made up of two protons and two neutrons quite easily. Such a nucleus, if it acquired its quota of electrons, would become an atom of helium, the same stuff that is used to fill children's balloons at carnivals. But to create more complex nuclei, you have to add more protons and neutrons to the helium. Now, if you add either particle to a helium nucleus, you get an isostope of helium (two protons, three neutrons), or an isotope of the next heaviest element, lithium (three protons, two neutrons). Both of these nuclei are unstable and come apart quickly. Consequently, the only way to create still more complex nuclei is to have the unstable nucleus incorporate something else before it comes apart, which is a pretty unlikely event.

It turns out that no matter how you try to build up complex nuclei from the stuff that can be made easily in the Big Bang, you run into the same sort of difficulty. As soon as you have nuclei with more than four particles, the instability of the new nucleus stops the building chain cold. The net result is that the Big Bang ended with the universe full of hydrogen and helium, a small amount of stable lithium (three protons, four neutrons), and very little else.

At this point, a moment's reflection will suggest that a world in which helium was the heaviest atom would be a pretty dull place. There would be no carbon to make living creatures, no iron for structure, no calcium for bones, no oxygen for water, and no silicon or aluminum or sodium to make rocks. We know, of course, that all of these elements exist in great quantities on the earth. If they couldn't have been synthesized in the Big Bang, then they must have been made at a later date in some other locale. Our attention, therefore, turns to the stars.

As we have already intimated, the energy that allows stars to shine is generated in nuclear reactions in the stellar core. They are fusion reactions—reactions in which two lighter nuclei come together to form a more complex one. In the process, energy is released in the form of energetic particles, and it is this energy that percolates out through the star and is eventually perceived by us as light and other forms of radiation. It is the pressure created by the particles streaming outward from the star's core that counteracts the force of gravity and allows the star to remain in a fairly stable state for billions of years.

In the sun, hydrogen nuclei (protons) interact in a complicated series of reactions whose end product is helium. Eventually, when all the hydrogen at the core has been "burned," the sun will start to collapse inward. This collapse will raise the temperature and pressure at the core, until the temperature gets so high that the electrical repulsion between the alpha particles is overcome. At this point, a new process can take place: three alphas can combine to make the nucleus of the carbon atom (six protons, six neutrons). The "ashes" of the old fire—helium—will become the fuel when the new fire is ignited. For the sun, this will be the end of the line. It simply isn't massive enough for further collapse to raise the core temperature high enough to ignite the next reaction. Other stars, however, can have this collapse-and-reignition process repeated many times, producing nuclei up to iron (twenty-six protons, thirty neutrons). Since iron is the most tightly bound of all nuclei, any reaction that changes it requires an input of energy. It is therefore unburnable—the ultimate nuclear ash.

So in stars, where high temperatures are maintained over billions of years, a lot of interesting elements can be synthesized. Unfortunately, in sedate stars like our sun these elements never get back out into the galaxy. The carbon core that will eventually form in our own sun, for example, will remain in the sun as it slowly dies and becomes a cosmic ember. In some stars, a small part of the elements created in the later phases of evolution will be returned to the interstellar medium through the solar wind; but by and large, what is made in these sorts of stars stays put. Thus, even though they can serve as crucibles for the formation of elements, they won't contribute much to the iron in your blood.

This leaves supernovae as the last resort in our search. The details of how a supernova functions are still being worked out by astrophysicists, but the general outline of events is pretty clear. A star more than ten times as massive as the sun quickly burns through all of the nuclear reactions required to form iron. When the temperature at the core is high enough for iron to be made, the lower temperature in a shell surrounding the core is still high enough for the synthesis of silicon, and in a shell surrounding that, the temperature is still high enough to make carbon—and so on. The star becomes an "onion," with different elements from helium to iron being created in layers as we move toward the center.

When the iron core gets to be about 1.4 times as massive as the sun, it can no longer be supported against the inward pull of gravity. The core of the giant star undergoes a massive and sudden collapse, going from something the size of the earth to something a few miles across in a matter of seconds. In the process, the internal pressure becomes so high that free electrons are forced into the iron nuclei, where they combine with the protons to form neutrons. The end result is an unimaginably dense collection of neutrons, the neutron star.

While this process is going on in the core, the outer shells have had the rug pulled out from under them. They start to fall in, meet the rebounding neutron core, and all hell breaks loose. The material is heated to the point where nuclei of every chemical element from helium through uranium (92 protons, 126 neutrons) are created. At the same time, the force unleashed in the collapse tear the star apart, spewing the newly created elements back into the interstellar medium where, as we have seen, they serve as the building blocks for new stars and solar systems.

In the region where our sun and the planets would someday form, gas that had been enriched by several of these supernova explosions collected. This gas formed the raw materials out of which the processes of gravitational collapse and accretion created the earth. Consequently, anyone seeking to answer the question "Where did all those rocks come from?" will ultimately come to the conclusion that the rocks and the mountains which they form came from a series of stellar explosions marking the deaths of giant stars. It's an interesting fact that the atoms that compose everything interesting around you—the earth, the sky, your own body—were fabricated in a series of events that took place in hardly more than a few hours.

# SIX

# The Great Shell Game

*Nothing can be made from nothing.*

—LUCRETIUS,
*The Nature of the Universe*

**M**ILLIONS OF AMERICANS have become familiar with Devil's Tower because of the role it played in *Close Encounters of the Third Kind* (a movie I have steadfastly refused to see). Towering 867 feet above the rolling rangeland of eastern Wyoming, the tower is an impressive sight, movie or no. The area was preserved as the nation's first National Monument by Theodore Roosevelt in 1906, and today serves as a mecca for tourists and climbers (See photo on page 87).

The most striking feature of Devil's Tower is the long vertical grooves that run up its sides. According to one story told by Kiowa medicine men, the grooves were made by a giant bear who was trying to get at a group of Indians on the top of the tower. Apparently, the wife of one of the Indians had been kidnapped by the Bear, and the Indian had stolen his wife back. After the bear had clawed the grooves in the tower, he was dispatched by a series of magic arrows. In another version of the story, the bear was trying to kill seven sisters who sought refuge in the stump of an old tree. The stump grew to become the tower and the sisters became the stars of the Big Dipper. These tales, like so much Native American lore, belong to the land from which they came—in this case, the open, arid environment of the high plains.

The modern explanation of the grooves of Devil's Tower is less dramatic but every bit as fascinating. Approximately 60 million years ago, an intrusion of magma from the earth's upper mantle forced its way into the thick layer of sedimentary rocks that form most of the landscape in the area. When its motive force was spent, the magma cooled, forming a kernel of hard igneous rock in the middle of the softer sandstone (see figure 6–1). Over the ages the sedimentary rock was weathered away, but the intrusion, being made of much tougher stuff, remained. Devil's Tower, then, is what is left once all the surrounding rock had worn away. I know of no better way to bring home the sheer magnitude of the geological process than to stand at the foot of the tower, look up, and realize that 60 million years ago you would have been buried under a thousand feet or more of solid rock.

Actually, the "claw marks" on the side of Devil's Tower were not made after the tower was exposed to the air; they were part of the structure from the time when the magma cooled in its underground chamber. In the molten state, the atoms in the magma were free to move throughout the fluid rock. When the intrusion had cooled for a while, the temperature dropped below the melting point and the atoms were locked into a crystalline

FIGURE 6.1

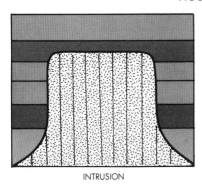 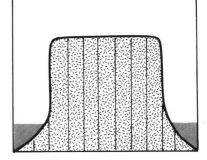

INTRUSION

structure. At this point the column was still hot, but atoms were no longer able to move around. As the heat continued to flow from the intrusion to its surroundings, the great mass began to shrink and contract. The contraction, in turn, set up stresses in the rock. In a large body like the Devil's Tower, the temperature rises uniformly as we move inward toward the center of the rock; so the rock along each plane parallel to the outer surface will have the same temperature. In this situation, it turns out that stress builds up inside the rock along these parallel planes, as shown in figure 6–2 below. Eventually, the stress builds up to the point where it can no longer be supported, and the rock cracks. If we apply the sort of knowledge of mineral structure that we will develop later in this chapter to the question of how the cracking occurs, detailed calculations say that cracks ought to radiate out from the lines of maximum stress at angles of 120°, as shown. The result is that the mass of rock, originally uniform, is split into a series of vertical hexagonal columns. The small triangular columns (shaded in figure 6–2) are no longer attached to the main mass of the rock, and quickly fall out. The result is a rock mass whose outer surface is marked with a series of long, vertical grooves.

FIGURE 6.2

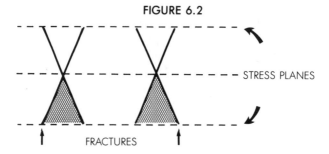

STRESS PLANES

FRACTURES

Geologists call this sort of thing "columnar jointing." It is a fairly common phenomenon, occurring in the Palisade Hills of New Jersey and New York as well as throughout the western United States. For some reason, the names attached to these formations tend to run to the diabolic—Devil's Tower, Devil's Postpile, and the like.

A number of interesting questions come to mind once we understand the "claw marks" on geological formations like the Devil's Tower. One is theoretical: Why should the magma split

A view of Devil's Tower, showing the tall columns shading into the jumbled rock at the base. Devil's Tower National Monument, Wyoming.

into hexagons rather than into other shapes like squares or triangles? A second is more empirical. If you look closely at the picture of Devil's Tower, you will note that the hexagonal columns stretch upward for most of the height of the column, but do not cover its entire length. As the columns approach the base, they twist and then disappear into a massive, jumbled piece of rock. From our discussion of the processes that led to the Tower's existence, we know that the rock in the base must be the same as the rock in the columns. Presumably the base went through the same cooling process as everything else. Why, then, does it look so different?

The conventional answer to the second question is that the base of the tower, being buried deeper than the rest, cooled more slowly. Consequently, it was placed under less stress than the columns, so that the fracture lines resulting from the release of that stress are more irregular. The best analogy is to think of a piece of wood. When it is subjected to a sudden, sharp stress by an ax or sawblade, wood splits cleanly along a straight line. When bent over your knee, however, it is subjected to a lower level of stress applied over a longer period of time. The stress is relieved a little at a time, and the result is a splintering, irregular break rather than a clean one. Presumably, the base of the tower fractured in a similar way, producing the observed irregular mass.

The existence of two different kinds of rocks in Devil's Tower touches on an important aspect of the physicist's way of looking at the world. Physicists (and especially theoretical physicists like me) tend to think in terms of a philosophy known as reductionism. Our training teaches us to break things down into their simplest components. It is the study of these components, be they molecules, atoms, nuclei, or quarks, that constitutes the frontier of physics at any given time. The unspoken assumption behind this method is that once you understand the components, you understand the system. Devil's Tower, on the other hand, is mute testimony to the fact that the situation may not be so simple. The two types of rocks, after all, are made from exactly the same assortment of atoms, but they are very different in character. Clearly, although the rocks at the top and at the base look different, they would have roughly the same properties if we tested them in the laboratory. But at many places in nature, two minerals made from exactly the same set of atoms have completely different properties. These examples point out an important truth about the limitations of reductionism: a physical system can sometimes be more than just the sum total of its parts. If we want to understand something, we not only have to know what its constituents are but what relationship those parts have to each other when they're all together.

A familiar example of the same situation is the difference between diamond and graphite. Both are made of the same stuff—pure carbon. Yet diamond is the hardest material known, while graphite crumbles so easily that it is used for the so-called lead in pencils. Graphite leaves a visible streak of broken pieces when you draw it across paper, something a diamond could never do. This example could be multiplied endlessly, but the lesson is clear. There is something more to rocks and minerals than the inventory of atoms.

The way I have finally come to think of the relation between rocks and their component atoms reminds me of an old carnival attraction called the shell game. In this game, the operator gathers a group of customers around a table on which is located a small object such as a pea and several shells turned upside down. The pea at first is shown to be under one shell, then the shells are shifted around and the pea moved from one shell to another until the whole thing is just a blur. The procedure is accompanied by a ritual chant: "Keep you eye on the mark, ladies and gentlemen, the hand is quicker than the eye." The customer is supposed to bet on his ability to tell which shell has the pea under it at the end. Many an unsuspecting visitor has learned to his sorrow that the hand is indeed quicker than the eye, at least when the hand belongs to the carney man and the eye to the overconfident observer.

The one useful lesson to be learned from the shell game is that if we let the pea play the role of an individual atom, and the various shells the role of the niches in minerals into which that atom can be placed, the shell game supplies a good analogy for the life of atoms at the earth's surface. Atoms may be included in a molten upwelling of magma and get locked into a particular mineral when the magma cools. Later, the atoms may be in a part of the rock that is weathered, included in sediment, and then subjected to heat and pressure to form a metamorphic rock. In this process, the atoms move from one shell (the igneous rock) to another (the metamorphic). Later, the rock containing the atoms may be subducted and melted, so that the atoms can start the whole cycle again, perhaps in a place in a completely different kind of rock. Like the pea in the shell game, a particular atom is always somewhere; but its exact location is constantly changing.

The connection between atoms and the rocks built from them is properly the domain of a field of knowledge called mineralogy. I have to confess that this is a field I've always tried to avoid. There is a great deal in it to discourage the uninitiated, not the least being a seemingly infinite collection of different names for different rocks (mostly ending in "ite"). Geologists started naming rocks long before anyone had any idea of atomic structure, or even of the existence of atoms. The names, therefore, tend to be idisosyncratic and peculiar. Sometimes they derive from the place where the mineral was first identified and studied— labradorite (Labrador), or franklinite (Franklin, New Jersey).

Sometimes they are named after people, either their discoverers or someone the discoverer wants to honor—goethite, for example, in honor of the poet Goethe. Geologists apparently still consider this to be a reasonable way to proceed, witness the lunar mineral armalcolite, named after the first team of Apollo astronauts to go to the moon (Armstrong, Aldrin, and Collins).

This sort of naming scheme makes for colorful lectures and books, but it creates difficulties for the neophyte rockhound or mineralogist, who is confronted with thousands of uncorrelated names to memorize. Sometimes the namers seemed to go to unusual lengths to create new types of rocks. I have been told, for example, that a number of mineral names refer to rocks found only in a single outcropping less than a foot square!

When it comes to understanding why a particular mineral has the properties it does, we have to move away from field geology and into the more arcane world of the atom. When various atoms are assembled into a given rock, the properties of that rock depend on two things: the kind of atoms in the assembly and the way that they are bound together and arranged. The binding and arranging, in turn, depend on the conditions that obtain at the time the rock is formed, particularly the conditions of temperature and pressure. We can examine each of these factors separately.

## Binding

If a rock holds together, obviously the atoms that make it up have some way of sticking together. There are four different methods by which atoms can bind themselves to one another (illustrated in figure 6–3 on the following page):

1. *ionic binding*—an electron from one atom jumps over to another, creating two ions of opposite charge, which then attract each other through the normal electrical force. (You will remember that an atom from which an electron has been removed or to which an electron has been added is called an ion.)

2. *covalent bonding*—two atoms share a pair of electrons.

3. *metallic bonding*—each atom in a given metal contributes an electron, which is then shared among the atoms as a whole.

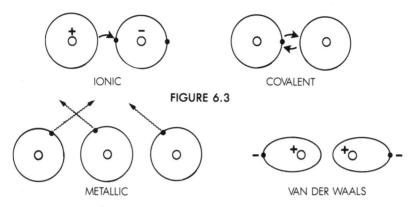

FIGURE 6.3

4. *van der Waals binding*—the electrons in each atom repel those in its neighbor, resulting in distorted atoms and a weak binding force.

The formation of a rock is no different from any other kind of formation by chemical reactions whereby atoms rearrange themselves to make new materials. As atoms approach each other, they first respond to the outer electrons in their neighbors, just as someone approaching the solar system would become aware of the outer planets before he saw the earth. Consequently, the chemical properties of any material are determined primarily by the number of electrons in the outer orbit of the atom, and not by the total number of electrons. That is why hydrogen (one electron) and sodium (eleven electrons, but only one in an outer orbit) have similar properties.

Let's take a simple example to demonstrate how binding in minerals works. Carbon atoms have six electrons each, of which four lie in the outermost orbit. Carbon atoms, then, have four outer electrons available for participation in the binding process. When these atoms are brought together under high pressure, they are packed together into a structure in which each carbon atom forms four bonds with other carbon atoms. The result is a three-dimensional structure, in which each carbon atom is locked tightly to its neighbors, as on the left in figure 6–4. In such a structure, it is hard to remove a single carbon atom from its place; so the resulting material is very hard. In fact, the material thus formed is diamond. The pressures needed to form diamonds are much higher than those normally available in the

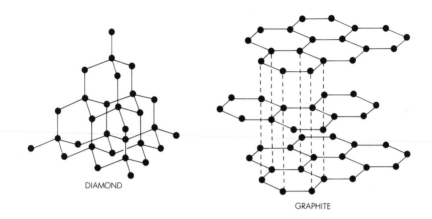

DIAMOND

GRAPHITE

**FIGURE 6.4**

earth's crust, so they have to be formed in the mantle and transported upward to the earth's crust. This explains why they are so rare.

At lower pressures, the carbon atoms are not forces so closely together and the resulting bonding pattern is different. In one such pattern, shown on the right in figure 6–4, carbon atoms form three covalent bonds with their neighbors in two-dimensional sheets, but form van der Waals bonds between atoms in neighboring sheets. This is the structure of graphite. Since van der Waals forces are relatively weak, it is easy for external forces to break the bonds between the sheets of carbon in this structure. This is why a pencil writes—the writing is composed of sheets that have been sloughed off when the van der Waals binding was broken.

The upshot of atoms' ability to form different bonds is that even in the simplest possible mineral, with only one sort of atom involved, it is possible to have very different physical properties depending on the way the atoms arrange themselves in the structure. Most rocks, of course, are much more complex than this. They are made up of combinations of many different types of atoms, so that in addition to thinking about the different types of binding forces between atoms, we have to think about the way that atoms of different sizes and different electron structure can be put together.

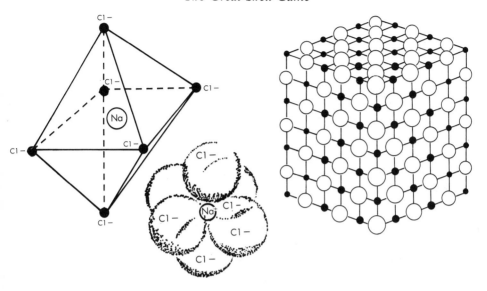

**FIGURE 6.5**

## *Arranging Atoms*

Most common minerals involve ionic binding, alone or in conjunction with one other type. Think of each ion in the mineral as an impenetrable billiard ball. The billiard balls will be of different sizes, of course, just as the atoms are. The general rule that tells us how the various atoms in a given structure will arrange themselves is simple: the mineral will be arranged so that as many of the billiard balls representing negative ions are touching each positive ion as possible.

Take common table salt as an example. It is, as you probably know, made up of equal parts of sodium and chlorine atoms. The sodium atoms give up one electron each to become the positive ions, while the chlorine atoms acquire one electron each to become the negative ions. The sodium atom is much smaller than that of chlorine, and the maximum number of chlorines that can touch the sodium atom is six, as shown on the left in figure 6–5 above. If we multiply this basic unit many times, we get a crystal structure like the one on the right, where sodium and chlorine ions alternate in such a way that every sodium atoms is surrounded by six chlorine neighbors.

Obviously, this arrangement results in the most compact pack-ing of the atoms in a grain of salt; but there is a deeper reason why this simple geometrical arrangement appears in nature. The fact of the matter is that the arrangement represents not only the most efficient packing but also the state of lowest energy (and thus greatest stability) of the sodium-chlorine system. In order to move any one of the negatively charged chlorine atoms away from the positively charged sodium, you will have to put energy into the system to overcome the electrical attraction be-tween the two ions. On the other hand, if you had a chlorine ion near a sodium ion with only five chlorines around it, the odd chlorine would fall toward the sodium, wedging other atom ions aside until it has fallen in as far as it could go. Just as a rock rolling down a hill demonstrates that every system in nature will tend toward the state of lowest energy, so too does the arrangement of atoms in minerals like table salt.

The same general pattern holds for other, more complex min-erals. Both silicon and oxygen, for example, are common ma-terials in rocks. Silicon usually gives up all four of its outer electrons to become an ion of charge $+4$, while oxygen attracts two electrons to become an ion of charge $-2$. Because oxygen is a large ion, only four oxygens can fit around a single silicon; so if there is plenty of oxygen around, we'll get a tetrahedral (four-surface) structure like the one shown in figure 6-6 below. Since four oxygens have a cumulative charge of $-8$, the total charge of the tetrahedron is $-4$ ($+4 + -8$), so more positive ions can be attached to it. Zircon, sometimes substituted for diamonds in jewelry, is a crystal whose basic building block is an ion of the element zirconium attached to the silicon-oxygen tetrahedron.

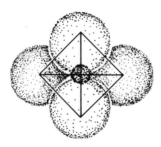

FIGURE 6.6

Sometimes minerals form in strange patterns. Here, water flowed into cracks in the large rock, carrying in atoms that eventually formed quartz. Beartooth Mountains, Montana.

If there isn't enough oxygen for each silicon to have its full share of four, then silicon atoms will share one or more of their oxygen atoms with their neighbors. Depending on the amount of sharing, you can get large sheets of interlocked silicon and oxygen (leading to a leaved structure in minerals like mica) or complicated three-dimensional interlocks (forming hard materials like quartz).

With more complicated minerals, made up of half a dozen or more difficult chemical elements, it's not hard to see that you could build up quite complex structures. Sometimes these structures will lead to hard materials, sometimes not. A type of clay called kaolinite, for example, is made up of aluminum, silicon, oxygen, and hydrogen. Ionic bonds form tightly bound sheets of material, but the sheets themselves are held together by the weak van der Waals forces. Like graphite, this clay is soft and easily broken up by finger pressure. Complexity does not, in and of itself, guarantee a strong material.

This Tinkertoy view of minerals also explains why it is that external factors like pressure can affect the structure of a material. The energy involved in ionic binding, as we saw in the case of table salt, depends on how many negative ions can surround a positive ion. With enough external pressure, you can distort and compress the ions and, in some cases, make it possible for them to rearrange themselves so that the positive and negative charges are closer together than they are normally. Once this rearrangement occurs, the electrical forces at the atomic level are usually strong enough to keep everything in place even after the external pressure has been removed. Sometimes, however, the atomic forces aren't equal to the job and the mineral slowly reverts to its original form. I have been told that this is true of diamonds, though the time it will take your jewelry to revert to graphite is probably as long as the lifetime of the earth. In other materials, once the change has occurred it is permanent. It's all a question of the way things stand at the atomic level.

### The Great Shell Game

The formation of the earth's rocks, then, always begins with the Tinkertoy assemblage of atoms in the crystals of the basic minerals. There are over three thousand of these arrangements known to geologists. Most of them, mercifully, are rare. Once the atoms arrange themselves in a given mineral, the igneous rock formed from the crystals begins to break down by the process of weathering, the resulting bits of rock are incorporated into sediment, and eventually make up the grains in sedimentary rocks. The sedimentary rock can then undergo metamorphosis (which we now understand to be nothing more than a shuffling of the basic atomic billiard balls under extreme pressure and temperature). At any stage the igneous, sedimentary, or metamorphic rock can be subducted, melted down, and returned to the magma from which it came. The entire cycle amounts to bringing the same atomic billiard balls together, shuffling them around, and letting them recombine to start the whole process over again.

But you don't have to take a grand tectonic view of the formation of minerals to see them as dynamic, rather than static, systems. From the Tinkertoy picture of the atomic building process, one point should be obvious. The only thing that matters when we ask whether a particular ion will fit into a particular spot in the structure is its size and its electrical charge. If two separate ions have the same size and charge, there is no reason why one can't be substituted for the other in a given mineral, a fact which has a number of consequences.

For one thing, it means that the number of different minerals that can be made from common materials is greatly expanded. Iron and magnesium ions, for example, are about the same size, so that one can be substituted for the other in a Tinkertoy structure. We would therefore expect to find minerals of the same structure formed at different places, but differing in the relative amounts of iron and magnesium they contain. Some of these minerals may have only iron, some only magnesium, some half of one, half of the other, and so on. This particular set of minerals is called the olivine group, and there are many other sets of substitutable ions like the iron–magnesium pair. The result is a great diversity of possible mineral types.

The possibility of substitution also means that the composition of a mineral need not be constant throughout its lifetime. If a member of the olivine group is exposed to magnesium ions in water (ions, for example, that have been leached from upstream rocks), the magnesium ions may replace some of the iron in the mineral. The result will be an iron (rather than a magnesium) ion flowing downstream and a magnesium (rather than an iron) ion in the mineral. Even on the relatively short time scales required for ionic substitution, minerals are dynamic, changing systems.

The substitution of one ion for another in minerals can have important consequences if one of those ions happens to be radioactive. Radioactivity is a property of the nucleus of an atom, but the fact that a nucleus will disintegrate through radioactive decay at some future time has no effect whatsoever on its electrons right now. As far as the mineral is concerned, a radioactive ion is just as good a piece of the Tinkertoy as any other; it will be incorporated into the mineral structure if it meets the size-and-charge test. The consequences can be both helpful and harmful.

One harmful consequence appeared during the 1950s, when many nations were conducting tests of nuclear weapons in the atmosphere. One product of these tests was strontium-90, a radioactive isotope of strontium. Now it happens that strontium and calcium are one of those pairs, like iron and magnesium, that replace each other in minerals. Strontium-90 in the fallout fell on grass, was eaten by cows and taken into the milk, where it replaced some of the calcium. When human beings drank the milk, these same strontium atoms could replace calcium in the bones. Once the strontium had been locked into place in the Tinkertoy of the bone structure, it remained in place until the nucleus disintegrated.

Half of the strontium-90 nuclei will decay in about twenty-eight years, which means that one-quarter of the original complement of ions will still be in place fifty-six years after they have taken their place in the bones. The combination of its ability to replace calcium coupled with a half-life comparable to the human lifespan makes strontium-90 one of the health problems associated with nuclear weapons. Its presence in fallout was certainly one of the factors that led to the treaty forbidding atmospheric tests of nuclear weapons.

But if the ability of one ion to replace another in minerals poses health hazards when strontium-90 is involved, it could turn out to be a boon when dealing with another problem connected with nuclear contamination: the problem of storing radioactive wastes. This question has received a good deal of publicity as experts have debated the environmental costs of nuclear power. The problem is itself unusual, not because it is difficult to find a solution, but because it is difficult to decide which of many possible solutions to implement. This has really confused engineers, who are so used to scrambling to find any workable solution that they tend to be nonplussed when confronted with a choice of equally efficient methods.

Ion substitution plays the leading role in many phases of the nuclear waste problem. Most schemes call for the waste to sit at the surface for a while so that the shorter-lived elements (like strontium-90) can decay. At that point the wastes are melted into some sort of structure like glass or cement, and the resulting pellets are surrounded by several layers of cement and metal before being packed into canisters. The canisters, in turn, are to be placed deep underground in some geological formation that has been stable and free from water for millions of years, and which, one hopes, will remain stable and dry for millions more. Thus separated from the biosphere, the nuclei of the long-lived radioactive elements decay according to their own internal clocks, causing no harm to living organisms.

This should work—unless things don't go according to plan. In the worst-case scenarios, engineers suppose that somehow water will enter the underground storage area, slowly leach away the protective metal and concrete, and eventually get at the waste itself. This would take a long time, of course; concrete, metal, and ceramics don't dissolve easily. Nevertheless, once radioactive material has entered the water, it can be carried by underground aquifers to areas where it might be drunk by human beings thousands of years in the future.

There are many lines of defense that can be erected to prevent this outcome. One is to build the storage facilities far from underground water. The second is to build the containers so that it takes water a long time to dissolve their shells. You can get some idea of what's involved here by asking how long it would take water to dissolve a glass jar. Design features buy time, time being all-important for the simple reason that nuclei in the waste are decaying all the while. With each passing year there is less dangerous material in the container.

The last line of defense is provided by nature. Suppose the worst happens and large numbers of radioactive nuclei enter the underground water system. This water seeps through the rock as it is transported from one place to another, perhaps moving at the rate of several hundred yards per year. If the radioactive atoms moved at this speed, ten or twenty thousand years would probably suffice to contaminate nearby water supplies. But as the atoms move through the rock, they undergo exchange reactions. For example, if the atom in question were strontium-90, then before it had moved very far it would replace an atom of calcium. The strontium-90 would then be locked into the minerals a mile underground while the harmless calcium would move downstream. This sort of process lengthens the time it would take water to carry the radioactive nuclei to human water supplies by values comparable to the decay times of the nuclei. So the exchange substitution reaction which makes the atoms dangerous when they enter the human body also makes it difficult for the atoms to get far from the storage area, even in the worst situation.

The Tinkertoy picture of minerals enables us to understand yet another approach to the problem of nuclear waste. There are many mineral structures in nature, and since the rules of their construction are understood, nothing prevents scientists from building others. All that we need to do is fit the right-sized atoms into the right places.

A number of scientists have pointed out that we could build minerals with structures tailormade to hold the various atoms that make up the radioactive waste. Many mineral structures that we know of have been stable over billions of years; if we duplicate these structures, substituting radioactive waste for some of the naturally occurring atoms, there is every reason to expect that the artificial minerals will be as stable as the natural ones. Thus by incorporating radioactive waste into minerals (instead of storing it in containers), we can guarantee that the atoms will stay where they are put.

The prospect is interesting because one point that critics of nuclear power often make is that nuclear wastes will have to be kept separate from the environment for hundreds of thousands of years—spans that are immensely long compared to the life of any human institution. But if the wastes are locked into stable mineral formations, we don't have to rely on human institutions to do the job. Nature will do it for us.

# SEVEN

# Old as the Hills

*How old are you, my pretty little miss?*
*How old are you, my honey?*
*She answered him with a tee-hee-hee*
*I'll be sixteen next Sunday*

*—"Blackjack Davey,"*
*English-American folk song*

HUMAN BEINGS are used to measuring time in seconds, days, and years. If we stray too far from this familiar temporal terrain, our intuition tends to get a little fuzzy. But nature is not bound by the human lifespan or reaction time, so we have had to learn to deal with time scales that are both so short and so long as to be virtually unimaginable. A list of some processes and their time scales is given in Table I.

Whenever we try to deal with things that take place on very long or very short scales, we have to develop special ways of measuring and quantifying time. In the world of the atoms and the atomic nucleus, the tools needed are fast electronics and huge particle accelerators.* The short-lived events have the advantage that they can be measured and analyzed within a single human lifetime. With the long-term events, this is not so. We

---

*A more detailed description of these techniques is given in my book *From Atoms to Quarks* (Scribners, 1980).

101

| TABLE I: TIME SCALES | |
|---|---|
| PROCESS | TIME |
| Reactions that hold the nucleus of an atom together | $10^{-24}$ sec. |
| Time for an electron to complete one orbit in an atom | $10^{-16}$ sec. |
| Time for one synapse in the brain to fire | .001 sec. |
| Human lifetime | 70 years |
| Written records | 5000 years |
| Lifetime of human race | 3 million years |
| Time for tectonic motion to replace the earth's ocean crust | 100 million years |
| Lifetime of the earth and solar system | 4.5 billion years |
| Age of universe | 15 billion years |

cannot, for example, watch the birth and death of a solar system, nor can we see the formation of a mountain chain from beginning to end. This means that we have to use the record left behind by past events to deduce both what happened and when it happened.

In the preceding chapters, we have bandied about a lot of numbers—the Rocky Mountains were born 65 million years ago, the oldest rocks are 3.8 billion years old and so on. When you go and look at a rock in the mountains, however, it doesn't carry a sign saying "65 million years old." It's just a rock. The problem of establishing a chronology—of attaching a sign to the rock that tells its age—has been and remains one of the central problems in the earth sciences.

In the eighteenth and nineteenth centuries geologists made a beginning at answering the question "How old?" by analyzing formations in the field. From our discussion of sedimentary rocks (see pp. 73–75), we know that the rocks in the lower layers of sediment must have been laid down before the rocks in the upper layers. This follows from our understanding of the sedimentation process, in which material rains down on the ocean floor over long periods of time: the stuff at the bottom had to be put there before the stuff on top. In sedimentary rocks, this establishes a relative time scale. If event A is deeper than event B, it happened earlier.

One problem that confronted early geologists was the fact that no single exposed bed of sediment covers the entire history of the earth. Even the Grand Canyon, one of the most impressive sedimentary beds ever to have been exposed for study, displays rocks created over a period of about 345 million years, beginning 570 million years ago, with the layers at the very top having been formed about 225 million years ago. In order to construct a continuous calendar throughout recorded time, it was necessary to compare sedimentary columns at different places. For example, it turns out that the rocks at Canyonlands National Park in Utah show exposed sedimentary rocks starting at 320 million years and ending 65 million years ago, while at Bryce Canyon National Park (also in Utah), the record starts 190 million years ago and comes almost to the present. These three parks, then, provide a series of overlapping yardsticks that can carry us back to a time 570 million years ago, an era that geologists call the end of the pre-Cambrian. The working out of the relative time scale didn't actually depend on these particular formations—they weren't even known to European geologists when some of the work was being done. They do, however, illustrate the method of matching different overlapping formations very nicely.

If you are of a skeptical frame of mind, you are probably wondering how, given my previous remarks, I can be so sure that these formations actually do overlap. How do I know, for example, that the layer of rock near the top of Grand Canyon was laid down at the same time as a layer near the bottom at Canyonlands? The answer to this question lies in the existence of markers in the rock, markers that have nothing to do with the geological processes itself. These markers are fossils—remains of skeletons of life forms long since vanished from the earth.

The establishment of the relative time scale, then, goes something like this: Two layers of sedimentary rocks in different locations are examined. If fossils of the same species of animals are found in both locations, then it is assumed that the rocks were laid down at the same time. This procedure allows us to match up pieces of different geological formations and arrive at a uniform relative time scale.

A few further points need to be made. In the first place, the example we are giving here, using three geological formations and only a few overlapping layers in each case, has been greatly simplified for ease of presentation. In reality, the time scales are put together by international committees of geologists who have the unenviable task of coordinating data from hundreds of different formations and making sure that everything fits together. The result of their work is a scale that allows us to say which formations are older or younger than others, regardless of their location.

Creationists sometimes argue that the geological time scale involves circular reasoning: fossils are used to date the rocks, and then rocks are used to date the fossils. It should be clear from the foregoing description that this argument is invalid, even for the relative time scale. The fossils are simply markers in the rocks. The only logical step needed to construct the relative time scale for fossils is to say that if fossil A occurs below fossil B in one formation, and fossil B occurs below fossil C in another, then fossil C was laid down after both A and B. There is no circularity here—only the normal rules of logic.

Finally, we should note that the time scale established in this way can only be used in formations in which large numbers of fossils are found. Consequently, the scale goes back only 570 million years, to the beginning of what geologists call the Cambrian period. Before that time there were few or no fossils, and the dating situation is rather murky.

Having made these points, you will note that I have been careful so far to use the term "relative time scale." The reason is quite simple. A relative time scale allows us to say that event A happened before event B, either because it appears below B in the same formation or because it appears below the time corresponding to B in a separate formation. But if we go further and ask "How long ago did A and B happen?", or "How much time intervened between A and B?", geological field studies can't help us. For example, if A and B are separated by six vertical feet of sandstone, this tells us nothing about the time of either event or the time between. Six feet of sandstone could easily have been laid down by a few major floods over a period of a few centuries, or it could have been laid down by a slow collection of sands over many millennia. There is no way to tell, just by looking at the sandstone, what its rate of formation was. There is no way, in other words, to attach numerical dates to either A, B, or the time that elapsed between them.

It was not until the early part of this century that it became possible to go beyond this sort of ordering process and attach real dates to the events we see in geological formations. The time scale formed in this way, with events numbered as well as ordered, is called the absolute or radiometric time scale. Establishing the absolute time scale still depends on geological field work, of course, but it also depends—critically—on the development of nuclear physics. In fact, I know of no better demonstration of the interconnectedness of scientific knowledge than the fact that studies of the atomic nucleus in laboratories around the world in the early twentieth century now enable us to say with some confidence that the last Tricerotops walked the plains of Montana 65 million years ago.

To understand how we use nuclear physics to unravel the earth's history, we have to make a short digression to recall what we learned about nuclei in chapter 5. The nuclei of atoms are made up of two major building blocks: protons (with a positive electrical charge) and neutrons (with no charge). The nucleus is about 100,000 times smaller than the atom itself. Most of the bulk of the atom is empty space, with a few electrons circling in orbit. You can get an idea of the relative sizes of these things by noting that if the nucleus of an oxygen atom were the size of a golf ball, then the rest of the atom would correspond to eight electrons, each smaller than a grain of sand, circling in orbits about the size of a large city.

Because atoms are electrically neutral, it follows that in the normal situation there are as many negatively charged electrons in orbit as there are positively charged protons in the nucleus. Thus, the identity of the atom depends on the number of protons in its nucleus. For example, the chemical properties of oxygen depend on the fact that there are eight electrons in orbit around the nucleus, but the number of electrons depends on the fact that there are eight protons in the nucleus. Should some process cause the number of protons in the nucleus to change, the atom will either lose electrons or pick up loose electrons from its surroundings to restore its electrical neutrality. It will, in other words, become a different atom with a different electron complement and, therefore, different chemical properties and *a different name*. For example, if we removed two protons from an oxygen nucleus, we would eventually wind up with an atom with six, rather than eight, electrons in orbit. This new atom would have the same chemical properties as any other atom with six electrons, and we would say that it was an atom of carbon.

For centuries, medieval alchemists searched for the philosopher's stone—the mystical substance that could change lead into gold. Unfortunately, the alchemists worked only with the chemical properties of materials, determined primarily by the electrons; they couldn't get at the nuclei of their atoms. In a sense, the particle accelerator, with its capability of adding and removing material from the nucleus, is the true philosopher's stone, capable of transmuting the elements themselves.

It is important to understand that it is the number of protons in the nucleus that governs the chemical properties of the atom, and not the number of neutrons. If we had extracted two neutrons, which are electrically neutral, from the nucleus of the oxygen atom in our example, we would not have changed the electron complement at all. There would still be eight protons and eight electrons, although there would be only six neutrons instead of the normal eight. The chemical properties of the atom would be unchanged, but its mass would be different. To a first approximation, all of the mass of an atom resides in the nucleus, and, since the masses of the proton and neutron are almost equal, it is customary to designate the mass by stating the number of protons and neutrons. Normal oxygen, with eight of each, is called oxygen-16, or $^{16}O$. With two neutrons removed, we would have oxygen-14, or $^{14}O$, while with two protons removed we would have carbon-14, or $^{14}C$. Two atoms that have the same number of protons but different numbers of neutrons are said to be isotopes of each other. Isotopes have identical chemical properties.

Most of the materials with which we are familiar are made up of stable atoms—that is, they are made of atoms whose nuclei do not spontaneously change their identity. There are substances in nature, however, for which this is not true. These atoms, of which uranium is the most familiar, have the property that the nucleus left to itself will, after a certain amount of time, emit charged particles, thereby changing the number of protons in the nucleus. For example, an atom of uranium-238 (92 protons, 146 neutrons) will emit two protons and two neutrons (an alpha particle), transforming itself into the nucleus of the element thorium-thorium-234, to be precise. We say that uranium is a radioactive substance, and that the uranium atom decays by emitting the two protons and two neutrons.

For the development of the geological time scale, the most important fact about radioactive nuclei is that each species decays according to its own internal clock. A collection of uranium-238 nuclei, for example, will begin decaying as soon as the nuclei are formed, but will do so very slowly. If we start with 1000 nuclei, then after 4.5 billion years there will be 500 left, after an additional 4.5 billion years there will be 250, and so on. The length of time it takes for half of the original complement of nuclei to decay is called the half-life of the nucleus. Therefore, for uranium-238 the half-life is 4.5 billion years.

Uranium-238 has one of the longest half-lives known; the half-lives of other radioactive nuclei range all the way from billions of years down to fractions of a microsecond. The elements that are most valuable for the geological time scale are those with long half-lives. We give below the principal elements we'll be using, together with their half-lives and the nuclei into which they decay.

| ELEMENT | HALF-LIFE (YRS) | DECAY PRODUCT |
|---------|-----------------|---------------|
| carbon-14 | 5730 | nitrogen-14 |
| potassium-40 | 1.3 billion | argon-40 |
| rubidium-87 | 47 billion | strontium-87 |
| uranium-235 | 710 million | lead-207 |
| uranium-238 | 4.5 billion | lead-206 |

The last two entries in the table represent not one decay, but rather a chain of decays that ultimately ends up as the stable isotope of lead indicated on the right.

In order to use our knowledge of radioactive nuclei to derive a time scale, we have to note two important facts:

1. Radioactive isotopes of an element can be incorporated into minerals and rocks in the same way as stable isotopes; and

2. Once incorporated, radioactive nuclei decay according to the dictates of their own internal clocks.

If we know how many atoms of a radioactive nuclei have been incorporated into a given rock when the rock was formed, and if we then measure the number of radioactive nuclei left today, we can discover how long it has been since the rock was made. The nuclear clocks thus permit us to assign absolute ages to geological formations, rather than merely relative ages.

Let's take the potassium-argon clock as an example of how the system works. Potassium is a fairly common element, and atoms of potassium appear in many different minerals; indeed, they make up over 2 percent of all igneous rocks on earth. In a normal mixture of potassium, 93 percent will be potassium-39 and only .01 percent potassium-40. Thus, the decay of the radioactive nucleus affects only a very small number of locations in the mineral structure and does not have much effect on the overall properties of the rock. What makes potassium-argon dating such a powerful tool is the fact that argon is almost never taken into a mineral structure (it is normally an inert, unreactive gas similar to helium). Consequently, if we start by counting a certain number of potassium-40 atoms and watch them decay, we can be sure that if we add the potassium-40 left over to the argon atoms at any time in the future, the total will be equal to the number of potassium-40 atoms with which we started. We can make this statement with confidence because we know that each decay converts one potassium nucleus into one argon nucleus, and every argon atom in the rock must be the result of such a decay.

Let's take a fictitious example. Suppose a rock is analyzed and found to contain 1000 atoms of potassium-40 and 1000 atoms of argon-40. The first thing we can say is that when the rock was formed, there were 2000 atoms of potassium-40 in it. We then note that exactly half of these have been converted into argon, from which we conclude that the mineral has been around for exactly one half-life of the potassium. From the table, we can see this means that the rock was formed 1.3 billion years ago. You can test your understanding of this argument by going through the above argument for a rock in which there are 500 atoms of potassium-40 and 1500 of argon-40. Do you get 2.6 billion years?

All the other radiometric dating techniques follow the same principle, although they don't all use exactly the same logical sequence to arrive at an answer. All, however, share certain limitations. In the first place, it is essential to the accuracy of the method that none of the product nuclei escape. Argon, as we have noted, is a gas, so if a rock is heated in the normal course of the geological cycle, some of the argon may be driven off. In effect, such an event resets the geological clock. When the rock is analyzed, the date determined by the radiometric techniques will be that of the heating event, not the formation of the rock. Sometimes this problem can be dealt with by analyzing an entire rock rather than just certain minerals in it, since the decay nuclei may well be in the rock even when they're no longer locked into the mineral structure. This sort of error leads to an age that is too small; it always underestimates the true age.

The second shortcoming of the radiometric dating technique is that it is very difficult to use it to date sedimentary rocks. The reason is this: Sedimentary rocks are made by the cementing together of grains weathered from older rocks. If we determined the radiometric age of a single grain of sandstone, it would tell us when the rock *from which that grain originated* was formed, but it would tell us nothing about when the grain was incorporated into the sandstone. Each grain in the sandstone might well come from a different rock, so radiometric dating would be useless. To establish true radiometric dates for sedimentary formations, we have to find igneous or metamorphic rocks from which we can infer the dates of the sediment. A layer of sedimentary rock sandwiched between lava flows, for example, would necessarily have been formed after the lower one and before the upper one.

All the dates I've used in this book are based on radiometric dating techniques. A number of different scales have been developed for assigning numbers to different geological events— I've counted at least eight in the literature. These scales are in broad agreement with each other, but differ in detail. The United States Geological Survey, for example, uses one scale, the British Geological Union a slightly different one. Still, if you keep in mind that the detailed number assigned to a given event may be uncertain at the level of 10–20 percent, you can use almost any time scale to talk about the past. Errors of this type, while important to scientists working in the field, make little difference to the overall view of the earth's history.

## The Age of the Earth

Estimating the age of the earth has long been a major intellectual pastime of the learned. In the seventeenth century the Anglo-Irish bishop Jame Ussher, by counting up the lives in biblical genealogy, concluded that the earth had been created on Tuesday, October 26, 4004 B.C., at nine o'clock in the morning. Later scientists used the saltiness of the ocean, the heat flow through the earth's crust, and the rate of formation of sediments in modern rivers to estimate the same number. Insofar as one can identify historical themes in such widely diverse fields of study, the trend throughout the nineteenth and early twentieth century was to assign longer and longer lifespans to our planet.

With the advent of radiometric dating, an important new tool was added to the equipment of those seeking to date the earth. The new techniques depend on a simple premise: The earth must be at least as old as the oldest rock on its surface. If we can find a rock that is of any datable age, we can be sure that the earth must be older than that age.

By 1931, at a meeting of the National Research Council in Washington, D.C., American geologist Arthur Holmes was able to state that the earth was at least 1.46 billion years old. This assertion was based on the uranium-lead dating of minerals taken from the Black Hills of South Dakota. From that time on, the location of the world's oldest rock jumped from continent to continent, residing for short periods of time in the USSR, South Africa, and the Congo. I was delighted to discover that for a short period in the 1960s, the world's oldest stone was a 3.1 billion-year-old zircon found in the Beartooth Mountains of Montana. It was coming to know these mountains that eventually led me to write this book.

Today, the world's oldest known rocks are in a formation of twisted gray metamorphic structure on the western coast of

Greenland. Dated by the rubidium-strontium method, these rocks are almost 3.8 billion years old. Since some of the rocks are clearly sedimentary in origin, this finding also indicates that the earth had oceans (or at least large bodies of water) on its surface 3.8 billion years ago.

The question that faces us, then, is how to go beyond the age of the oldest known rocks to the age of the earth itself. If you think back to the discussion of the earth's formation in chapter 4, you will realize that radiometric dating cannot be used to determine the age of anything that happened before the melting and differentiation of the earth. In effect, the melting reset all the geological clocks on earth. If we want to go beyond that event to the formation of the earth itself, we will have to use an indirect method of arriving at an estimate.

One such estimate is based on the radiometric dating of meteorites. Both uranium-lead and rubidium-strontium measurements indicate that meteorites striking the earth were formed 4.6 billion years ago. If we believe that these meteorites represent material left over from the construction of the planets, then we say that the earth must have formed at the same time they did—4.6 billion years ago.

A second line of inference is to look at the dating of rocks taken from the moon. Again, we find that standard techniques (primarily potassium-argon) give an age for moon rocks of about 4.6 billion years. The moon, as we saw, underwent differentiation only at the surface, and solidified quickly. Consequently, geological clocks on its surface were reset much earlier than those on the earth, and their age is a much better indication of the time of the moon's formation than are the corresponding ages for earth rocks. If you think the earth and the moon formed at the same time, then you would conclude that the earth is 4.6 billion years old.

We can get still another estimate of the earth's lifetime by looking at isotopes of lead. Lead has a number of stable nuclei, but we will consider just three—the ones numbered 204, 206, and 207. Lead-204 is not the end product of any radioactive decay chains, so the amount of lead-204 will remain the same throughout time. Lead-206 and 207, however, are the end products of the decays of uranium (see the table on p. 107); consequently as time goes on, the amount of both of these elements will increase in any mineral that contains uranium.

Occasionally, we can find a meteorite that contains no uranium-bearing minerals. The body that caused the great crater in Arizona was made of such minerals, for example. In these meteorites, the relative numbers of lead-204, -206, and -207 should be the same as they were in the primordial cloud from which the solar system formed. Since the earth was formed from the same cloud, the relative abundances of these three isotopes must have been the same when the earth was formed as it is in the meteorites today.

Since the formation of the earth, the amount of lead-206 has been augmented by the decay of uranium-238, and the amount of lead-207 has been augmented at a different rate by the decay of uranium-235. This means that the total inventory of these two isotopes on the earth has been changing steadily since the creation, so that minerals containing lead (but no uranium) formed at different times will have different amounts of lead-206 and -207, depending on how much of each isotope was present when that particular mineral was formed. If we simply look at a number of large lead deposits on the earth and count the isotopes present, we find that the isotope abundances match exactly the predicted curve. We also find that the time it would have taken for an initial abundance like that found in the meteorites to turn into those now found on earth is 4.6 billion years. This is another good estimate of the age of the earth.

It's important to understand the logic of the procedure outlined here. The assumptions underlying any single method of determining the earth's age may be questioned, so a single determination may be regarded with some skepticism. When several unrelated estimates all yield the same result, however, the confidence you can have in the conclusion goes up considerably.

I have described four ways of estimating the age of the earth:

| METHOD | AGE (Billion Years) |
|---|---|
| Dating of oldest terrestrial rocks | more than 3.8 |
| Dating of meteorites | 4.6 |
| Dating of oldest lunar samples | 4.6 |
| Counting lead isotope abundances | 4.6 |

It is hard to avoid the conclusion that our planet was formed 4.6 billion years ago.

## The Age of the Universe

The same sort of situation obtains when we talk about determining the age of the universe. There are several ways of determining its age, each open to question in and of itself. Once again, however, all three methods give approximately the same result, so that we can have some confidence in the final number. Unlike the case for the earth, however, our knowledge of the universe still appears to be uncertain by about a factor of 2.

One way of dating the universe is analogous to the technique of finding the oldest rock on the earth: Find the oldest star. Obviously, the age of this star must be less than the age of the universe, just as the oldest rock has to be younger than the earth. In chapter 5, we outlined the life cycle of a star, pointing out that all chemical elements heavier than helium are manufactured in stars and returned to the interstellar medium upon the star's death, there to be taken up into new stars. It follows that the amount of heavy elements in stars must have been increasing as the universe grows older and more of these elements are created in the stellar furnaces. It also follows that the earliest stars must be deficient in heavy elements compared to later stars. As a matter of fact, astronomers know that there is a large class of stars that are poor in heavy elements—they are called Population II stars. It is from among such stars that we expect to find the oldest members of the stellar family.

In the early 1970s, astronomers Icko Iben and Robert Rood, both then at MIT, studied clusters of old stars. By taking account of the amount of hydrogen that had been converted to helium in these stars, they were able to produce estimates of the amount of time the stars had been burning. These estimates depend, of course, on the amount of helium that had been incorporated into the stars initially; the more helium at the start, the less time it would take to bring the helium content to present levels. Models of helium production during the Big Bang generally predict that the universe started off with about 22 to 26 percent of its atoms in the form of helium, the rest in hydrogen. From these estimates, the results of Iben and Rood give the ages of the oldest stars as somewhere between 12 and 18 billion years, the most likely age being about 14 billion. Since stars form very quickly compared to the life of the universe, this result

*113*

would suggest an age for the universe of around 15 billion years.

A second way of estimating the lifetime of the universe is to use a technique very much like radiometric dating. In chapter 5, we saw that elements like uranium are produced in the final catastrophic stages of a supernova. While it may seem strange to talk about understanding what goes on in such a maelstrom, you have to remember that the reactions by which protons and neutrons are added to existing nuclei to make heavier nuclei have been studied in laboratories by nuclear physicists for decades; in fact, William Fowler of Cal. Tech. was awarded the Nobel Prize in 1983 in recognition of his many years of work in this field. From laboratory data, it is possible to calculate the relative amounts of various heavy elements produced in supernovae.

Once these elements are produced, many of them will start to decay. By measuring the relative abundances today and comparing them with what was produced in the supernovae, we can determine how long it has been since the radioactive decay processes started. For example the calculations tell us that in supernovae, uranium-238 will be produced less copiously than thorium-232. In fact, the calculations say that the ratio of thorium-232 to uranium-238 is 1.6 to 1. Now thorium-232 has a half-life of 14 billion years, while uranium-238, as we know, has a half-life of 4.5 billion years. Consequently, the uranium will disappear from a sample faster than the thorium, and as time goes by, the thorium will dominate more and more. Analysis of lunar rock samples shows the present ratio between the two elements to have climbed to 4.1 to 1 from its initial value of 1.6 to 1. It is then a simple matter to show that the material in that sample was created about 10 billion years ago. If all the heavy elements that made up the solar system were created in a single event, then this number would tell us that that event occurred 10 billion years ago, or 5.4 billion years before the creation of the earth.

It is very unlikely, however, that all the heavy elements on earth were created in a single supernova. There have been around a billion supernovae in the course of galactic history, and the present mix of elements on earth results from the output of a large number of them. When you take this into account it appears that the first supernova which contributed to our inventory of elements occurred at a time stretching back twice as long before the formation of the earth as the period calculated if you assume all the elements came from a single supernova. This means that the uranium-thorium ratio gives $2 \times 5.4 + 4.6 = 15.4$ billion years as the age of the first supernova. Since the lifetime of a large star is very short, this last figure is a pretty good estimate of the age of the universe.

In addition to uranium-thorium dates, you can use other isotope pairs to yield similar estimates—rhenium-187 and osmium-187 are another popular pair. When this is done and the uncertainties in various measurements are taken into account, the radiometric dating of the unverse once again leads to this result: The universe is between 8 and 19 billion years old, with the most likely age being 15 billion years.

The final estimate of the lifetime of the universe has to do with the fact that when we look at neighboring galaxies, we find that they are receding from us, with the more distant galaxies receding more rapidly than the closer ones. This tells us that the universe is expanding.* Now, if we know how the expansion is proceeding at present, it is very tempting to "run the film backward" and go back to a time when all the matter in the universe was crammed into a single point in space. The event in which this single point began to evolve into the present universe is called the Big Bang, and the date of the Big Bang is what is meant by the age of the universe.

The usual age quoted for the age of the universe from studies of the universal expansion is 15 to 20 billion years, but recently a rather violent scientific storm has broken out regarding this particular number. A group at the University of Texas at Austin has published its own value for the expansion rate, which corresponds to an age around 8 billion years and an upper limit of 12 billion. The debate has become one of those that periodically arouse strong passions among experts. This one has to do with arcane technical points in the analysis of the luminosity of galaxies at the very edge of the universe. My reading of the situation is that the astronomical community has taken note of the arguments advanced by the Texas group, but isn't giving up the conventional value of 15 billion years.

---

*A description of the expansion of the universe and the fascinating story of the early stages of the Big Bang can be found in Steven Weinberg's excellent book *The First Three Minutes* (Basic Books, 1977), or in my book *The Moment of Creation* (Scribners, 1983).

We can sum up the determination of the age of the universe as follows:

| METHOD | MOST PROBABLE AGE (billion years) | SPREAD (billion years) |
|---|---|---|
| Dating of oldest stars | 14 | 12 to 18 |
| Ratios of radioactive isotopes | 15 | 8 to 19 |
| Universal expansion | 15 | 15 to 20 |

The range in the last entry does not reflect the new uncertainty arising from the controversy mentioned above.

In summary, then, three independent measures give roughly the same result: The universe is about 15 billion years old and there is an uncertainty of about 5 billion years in either direction. Considering the difficulty of the problem and its philosophical importance, getting a number as close as this is a significant achievement.

## A Final Comment

People who call themselves creation scientists frequently seize on controversies like those surrounding the rate of universal expansion to argue that the earth was, in fact, created about the time Bishop Ussher said it was. At least they argue that the age of the earth and the universe is to be measured in thousands, rather than billions, of years. Such arguments completely ignore the fact that neither age is fixed by a single method of calculation or a single number. Rather, they are fixed by several independent analyses, which converge on one result.

Dissenters also ignore the fact that even in the most controversial case, the choice is between a universe a little less than 10 billion years old and a universe 15 billion years old. Neither party to the controversy would countenance a universe whose age was measured in millions (much less in thousands) of years. In this case, I'm afraid, the creationists are like someone who overhears an argument about whether a particular office is on the fifty-fifth or fifty-sixth floor of the Empire State Building and concludes that the building is only one story high.

# EIGHT

# Solid as a Rock

*The first and basic principle of all things is
water.*

—THALES OF MILETUS

*Water is an extraordinary substance, anomalous
in nearly all of its properties and easily the most
complex of all familiar substances....*

—*Encyclopaedia Britannica*

I T'S HARD TO WALK very far in the mountains—or elsewhere, for that matter—without encountering a scene
like the one shown on page 118. A large rock, once
whole, is broken into smaller pieces. We all learned in grade
school that this happens because water flows into cracks in the
rock, freezes, and expands. The force of the expansion widens
the crack, more water flows in, and the cycle is repeated until
the rock splits.

Geologists recognize this as one of the important ingredients
of weathering—the process by which rocks are converted into
soil. The broken rocks are a token of those slow, inexorable
processes that grind even the highest mountains down to hand-
fuls of loose soil and make way for the next set of mountains
to be raised.

There are other ways for rocks to be broken up, of course.
Anyone who has walked on sidewalks in old areas of town has

A large rock in the process of being broken up by the freezing of water in its cracks. Beartooth Mountains, Montana.

probably seen places where tree roots have cracked the cement. The same thing happens in the mountains—plants take root in crevices in a rock and the roots move down into small openings they find there. As they grow, they split the rock apart just as the ice does.

There are also chemical reactions that can weather rocks. Perhaps the most familiar of these is the rusting of iron. Water and oxygen from the air combine with iron to form a reddish, crumbly mineral called limonite—the stuff we usually call rust. You often see rocks with colored streaks down their sides where the native iron has been leached out by rusting. You can also see streaks of different colors, where analogous processes have acted on other metals.

Most of the activities that go on in weathering seem pretty straightforward. There are many examples of the same processes at work in our everyday life, most notably in weathering by the formation of ice. This last process depends on the fact that the ice formed when water freezes occupies a volume approximately 9 percent greater than that occupied by the water. If water is put into a closed container and frozen, it exerts a force great enough to break the container.

*118*

This property of ice often leads to major problems in water systems during cold weather. Everyone knows that a pipe allowed to freeze while full of water will break, or at least pop loose at the joints. Then, when the temperature goes up again, the pipe leaks. I hate to think of how many joints I've had to resolder because I was caught unprepared by the first freeze in autumn.

If you think about this common phenomenon for a moment, you'll see that if ice weren't less dense than water, life on earth could very well be impossible. To take a simple example, when a lake freezes in the winter, a layer of ice forms across the top. This layer has the effect of shielding the deeper part of the lake from the cold atmosphere, so that life can go on in the water. If, as with most other liquids, solid water were more dense than liquid, each layer of ice would sink to the bottom as it formed and the lake would freeze clear through, killing most of the living organisms. So the next time you see an ice cube floating in your Coke, you might reflect upon the thought that if the cube didn't float, aquatic life on earth would have to be confined to the tropics—if it existed at all.

The expansion of water when it freezes, then, is an aspect of nature that reaches into every part of our lives. This makes it all the more surprising that until quite recently, scientists didn't have a very clear idea of why it happens.

All matter exists in three phases: solid, liquid, and gas. In water, these phases correspond to ice, liquid water, and steam. A simplified picture of the three phases of matter is given in figure 8–1. In the solid phase, the atoms are locked together into the kind of Tinkertoy structure we discussed in chapter 6. Each atom is locked firmly to all the others in a lattice. If you push on one, it will not move unless all the others move as well. It is the existence of these internal forces that allows solids to

| SOLID | LIQUID | GAS |
|-------|--------|-----|

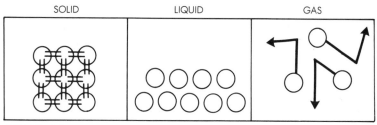

FIGURE 8.1

119

keep their shape and to resist forces that act on them. In general, atoms in solids are packed tightly together.

In a gas, on the other hand, no forces hold the atoms in place. They are free to move around and, in general, can move great distances before colliding with other atoms. The liquid state is somewhere between that of the solid and of gas. The atoms can move around, but they tend to be bound together by very weak forces. You can think of a liquid as being something like a bag of marbles: each marble can move by sliding over its fellows, but it requires some effort to overcome the friction between them.

The transition between the states of matter, which physicists call phase changes, are easy to understand when you have this simple picture of the atomic arrangements in mind. If you start with a solid and begin adding heat, you can imagine the atoms beginning to vibrate more and more vigorously. The atoms remain more or less in place, but they make larger and larger excursions from their home sites. Eventually, a point is reached where the vibrations become so violent that the interatomic forces can no longer keep the atoms in place. They tear loose from the lattice and begin to move about freely. The material changes from a solid to a liquid—and we say that melting has occurred. Freezing, the process of going from a liquid to a solid, is just melting in reverse.

If we continue to add heat to the liquid, the atoms will start to move with a higher and higher velocity. An atom near the surface of a liquid is normally kept in the liquid by the weak interatomic attraction that holds the liquid together; but if it approaches the surface with enough energy, it can break through and escape. In a sense, the atom is like the payload of a rocket launched from earth. If the rocket payload has enough energy, it can escape from the pull of the earth and move into space. An atom of liquid moves into the atmosphere in the same way. When the temperature is high enough, all of the atoms of the liquid will have enough energy to escape, and the liquid becomes a gas. We say it boils when it turns to steam.

With this background, we can return to our original question concerning the increase in volume when water freezes. Normally, when a liquid solidifies, the atoms within it become more efficiently packed. This results in a decrease in volume, a decrease that is typically on the order of 10 percent. The best way to understand this decrease is to think of packing a suitcase. If you just throw everything in, you will need more space for your things than if you packed everything neatly. The liquid is like a hastily packed suitcase—the atoms are just thrown in every which way. When the liquid freezes, the suitcase repacks itself in an orderly way and the atoms therefore take up less room; the solid has a smaller volume than the liquid.

Except for water. It isn't only this volume change that makes water an unusual fluid; water is different from other liquids in almost all of its physical properties. For example, from our discussion of the mechanism by which liquids boil, we would expect that the lighter the atoms in a liquid are, the easier it should be to give them enough energy to escape from the fluid. It is, after all, easier to launch a small satellite than the space shuttle. And if we look at most liquids, we find this general correspondence between atomic weights and boiling point—the higher the weight, the higher the temperature at which the liquid boils.

Except for water. If we used the results from other liquids to estimate the boiling point of water, we would predict that it would boil at around 135° below zero ($-93°C$) and freeze into a solid a few degrees below that. Yet here we are, happily swimming in oceans whose temperatures hover at 70° *above* zero without showing any signs whatsoever of boiling.

Similarly, if you think of the connection between temperature and the motion of atoms, you would expect that as temperatures rise, a fluid should expand. This follows from the fact that at high temperatures each atom in the fluid is moving faster, so that it carves out more elbow room for itself in its collisions with its neighbors. The extra elbow room, added up over all the atoms, results in a larger volume for the liquid as a whole. In most liquids, there is a general correspondence between temperature and volume.

Except for water. One of the most astonishing facts about water is that once it has melted, raising its temperature actually causes it to shrink until it warms up to about 40° Fahrenheit ($4°C$), at which point it starts expanding again. Why do water molecules give up their elbow room for a while, then change their minds and start regaining it again?

We could go on listing the anomalous properties of water for a long time, but let me finish with just one more. The specific heat of any substance is defined to be the amount of heat needed to raise the temperature of a given amount of the material one degree. Since temperature is a measure of the speed with which atoms in a material are moving, the specific heat tells us how much of the heat that enters the sample is converted into atomic motion. The higher the specific heat, the more heat you have to pump in to get a desired rise in temperature. Conversely, something with high specific heat will release more heat as it cools off, because it took more heat to raise its temperature in the first place.

When you add heat to a solid, you have to do two things: you have to make the atoms vibrate faster in their places, as we mentioned above, but you also have to stretch the Tinkertoy structure a little bit to accommodate the solid's expansion. Some of the energy put into a solid, then, goes into things other than atomic motion. In a liquid, where the interatomic forces are weak, not much energy has to go into the other things. Consequently, heat introduced into a solid will produce less atomic motion (and hence a lower final temperature) than will the same amount of heat introduced into a liquid. As you might expect from this argument, most substances have a higher heat capacity in their solid than in their liquid phase.

Except for water. The heat capacity of water is twice that of ice, and is, in fact, one of the highest known in nature. Somehow, only a part of the energy put into liquid water goes into atomic motion. The rest goes somewhere, but where?

The high heat capacity of water has many consequences. It explains why water is so often used as a heat reservoir in solar heating systems (see chapter 4). For each degree of temperature rise, it stores more energy in the form of heat than almost any other material. Another place where the high heat capacity of water plays an important role in life is in the great ocean currents like the Gulf Stream. Warmed in the tropics, water in these currents slowly rolls toward the poles, pumping heat into the atmosphere as it cools. The amounts of heat involved can be truly staggering. It has been estimated, for example, that the heat surrendered by the Gulf Stream in two hours exceeds what you could produce by burning the entire annual stock of coal mined in the world. The climatic effects of the ocean currents are well known. What isn't so well known, perhaps, is that these effects depend on the anomalous behavior of the heat capacity of water.

What I find startling about the state of knowledge about water is not that we know so little about it, but that there has been such a dearth of serious studies of water throughout history. I suppose it is a case of familiarity breeding contempt. The anomalies of water are so well known to us that they do not excite interest—at least until you start thinking about them. That we know so little about the most important liquid in our environment probably tells us a good deal about human psychology.

Throughout most of recorded history, water was considered one of the basic, indivisible elements from which the universe is made. In the sixth century B.C. the Greek philosopher and statesman Thales of Miletus argued that everything in the world is made of water. I suspect that he was led to this conclusion by the fact that water, alone among familiar materials, exists in all three phases of matter. If everything is either a solid, a liquid, or a gas, and if water can be all three, then it's not hard to see that water must be accorded a special place in the scheme of things. It was Thales' student, Anaximander, who added the elements earth, fire, and air to his teacher's list, thereby creating the familiar quartet of elements that survived well into our modern era.

The next significant step in the study of water didn't occur until the latter part of the eighteenth century, when chemists showed that water was not an element, but was made up of two other substances (hydrogen and oxygen) that were. The coining of the word "hydrogen" ("generator of water") reflects the new understanding of the composition of water. The German word *Wasserstoff* ("the stuff of water") makes the same point.

In modern terms, these early chemists found that the water molecule was made up of two atoms of hydrogen and one of oxygen. The representation of water as $H_2O$, which rests on this fact, is so familiar that it can honestly be said to be part of our folklore.

All the more amazing, then, that over a century passed from this discovery before any further thought was given to the nature of water. It wasn't until the 1890s that anyone looked at water as a subject for serious scientific work. At that time a number of people, including William Roentgen (the discover of X-rays), began to wonder about the odd behavior of water near the freezing point. Roentgen eventually developed a model in which he postulated that water above the freezing point had within it microscopic portions in which the molecules still retained the crystalline structure they have in ice. As water is heated, this remnant ice melts slowly; and as it does so, the volume continues to shrink. (This behavior arises because ice takes up more space than the same mass of water.) The shrinking process, so Roentgen thought, continues until all the remnant ice is gone, at which point additional heat causes the water to expand.

After Roentgen, it wasn't until the 1950s and 1960s that detailed models for the structure of water were developed. The most interesting of these bear a family resemblance to Roentgen's, but are much more detailed. The most interesting question to me, however, is not so much the details of our present understanding of water, but the fact that scientists waited so long to come to a serious study of its properties.

The attitude of most people is that once you know that water is good old $H_2O$, there's nothing more to be said. This attitude was well represented by an anecdote told by the British chemist Felix Franks.* One of the world's experts in the study of water, Franks was returning by train to his laboratory. Sharing his compartment was a student who had been interviewed for a job with the company Franks had been visiting. They struck up a conversation, and when Franks said his field of research was the study of water, the student "looked at me as though I were a lunatic and in a condescending manner informed me that water was just $H_2O$ and that anything one needed to know about it could be found on half a page of a standard textbook. The clear message was that I was wasting my life." A courtly gentleman of the old school, Franks refrained from comment at the time. He couldn't help adding, however, that he later learned the student hadn't gotten the job.

To study the properties of any material, you must ultimately study its atoms and the way they are assembled. Water is no

*Franks's marvelous book *Polywater* (MIT Press, 1981) is well worth reading if you are interested either in water or the way that scientists work.

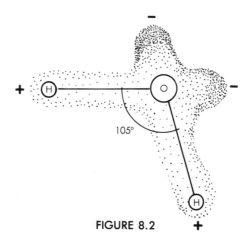

FIGURE 8.2

exception to this rule. The hydrogen and oxygen in water are held together by covalent bonds—bonds in which atoms are held together by the sharing of electrons. In such a situation, the electrons are not locked into a single place in the atom, but move around. We can illustrate this by representing the outer (shared) electrons in water by a cloud. When we work out the configuration of nuclei and electrons that leads to the lowest energy of the water molecule, we find that the water molecule is arranged as shown in figure 8–2. There is an angle of 105° between the lines drawn from the oxygen to the two hydrogens. The electrons spread out roughly in the X-shaped cloud depicted.

The electrons spend relatively little time around the hydrogen nuclei, which means that that end of the atom has, on the average, a net positive electrical charge. On the other hand, the electrons spend a fair amount of time out in the two lobes on the other side of the oxygen atom, giving that end of the molecule a peculiarly shaped negatively charged region. It is this odd and irregular arrangement of the charge in the water molecule that is responsible for all the anomalous properties of water.

The fact that the electrical charge in a water molecule is permanently distorted means that when the molecules come together in either a liquid or a solid, they will tend to line themselves up in a definite geometrical relationship, rather than

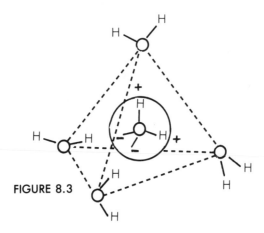

FIGURE 8.3

lining up randomly. We would expect that the negative part of one molecule would tend to line up next to the positive part of another; and indeed, this is what happens. As was the case for mineral structure, however (see chapter 6), this general rule has to be applied subject to the requirement that the molecules of water must be fitted together so that they do not overlap one another. The arrangement that satisfies both of these requirements for water molecules is shown in figure 8–3. This figure, known as a tetrahedron, has the property that the angle between the oxygen molecules in different molecules is 120°. (Don't confuse this angle with the 105° angle inside a single molecule.) The electrical force between the positive part of one molecule and the negative part of another is called the hydrogen bond. It is a relatively weak binding force.

When water freezes into ice, the tetrahedral structure gets locked in, and we have the molecules arranged as shown in figure 8–4. On a given sheet, there is an open hexagonal structure as shown on the left, and the sheets are locked together in a set of tetrahedra as shown on the right.

In ordinary ice, the hexagonal structure determines the properties of the solid. It is because of this structure, for example, that snowflakes have their familiar six-sided shape. It also explains why ice is less dense than water. The molecules are arranged in an open structure, with a great deal of free space in the hexagons. Obviously, fewer molecules can be packed into this structure than could be placed into something less open.

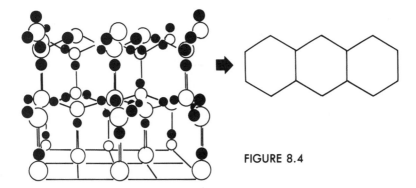

FIGURE 8.4

To use our suitcase analogy, the structure of ice is such that putting molecules into ice is like packing a suitcase in which every item is wrapped in sheets of paper so that no item can be crushed against another. If we require this extra packing material, it's clear we will be able to get fewer things into each suitcase. The open hexagonal structure of ice, in effect, requires that each molecule we put in be surrounded by a lot of open space. The same set of water molecules, then, will take up more space after the molecules have formed into ice than it did before. Consequently, water will expand when it freezes, and decrease in volume when it melts. The reason it does so is ultimately due to the way the electrical charge is arranged in the individual molecules.

But the tetrahedral structure of the water molecules doesn't disappear when ice melts. The "normal" picture of a liquid, in which molecules move randomly with respect to each other, doesn't apply to liquid water. The same electrical forces that lock the molecules together in ice operate between the molecules in a liquid. At temperatures above 32° Fahrenheit, these forces are not strong enough to overcome the normal agitation of the molecules, but the forces are still there. What happens is that in water, a certain percentage of the atoms tend to line up in tetrahedral structures, but these structures do not last very long. If we could take a snapshot of the molecules in water at a given instant, we would see most of the molecules in tetrahedra. If we took a picture a moment later, we would see the same thing.

If we followed an individual molecule, however, we would see that it was flitting from one tetrahedron to another between photographs. Thus, although individual molecules are free to move about in the liquid, the intermolecular electrical forces require them to occupy the appropriate spots in vacant places in tetrahedra most of the time.

This picture of water shows that most of its anomalous properties arise from the fact that, while it isn't a normal solid, it isn't really a liquid either. In a solid, the same atoms or molecules are always locked into the same place in a structure. In a normal liquid, there is no structure at all, and the molecules are free to move without regard for each other. Water is obviously somewhere between these two—the molecules are free to move, as in a liquid, but they are constrained to move in such a way as to preserve a solid-like structure. I like to express this by saying that *water never quite forgets that it was once ice.*

As Roentgen guessed, the anomalous properties of liquid water have to do with its icelike structure. For example, near the freezing point the fleeting structure of the liquid approaches closer and closer to the open hexagon of ice, causing the water to expand as it is cooled. What governs this process is not the collision between molecules, but the changes in the internal structure of the liquid. Water's other anomalous properties have a similar explanation; the specific heat, for example, is explained by the energy that has to be supplied to change the internal structure as the temperature increases.

The realization that water has a structure is fairly recent—it has been brought to a well-thought-out state only in the last quarter century. But though it is now proper to say that we understand the general features of liquid water, many of the details of its behavior as a liquid and as ice are still the subject of vigorous research and debate in the scientific community. The relative lack of understanding of the most important liquid in our lives was illustrated in a major scientific episode of the late 1960s and early 1970s, the famous (some would say infamous) case of polywater.

In 1962, a Russian chemist named Nikolai Fedyakin, working in the provincial town of Kostroma, reported a rather strange experimental result. He noted that when water was sealed into narrow tubes for several weeks, a thin layer of liquid grew on top of the water column. It seemed that pure water had separated into two kinds of liquids, and that what had been discovered, quite by accident, was a new form of water, a form that was eventually named "polywater."*

The history of this discovery is interesting because it illustrates many aspects of the way the scientific community works. The first thing that happened was that Fedyakin's result was totally ignored in the West. The reason for this is simple: even though Soviet scientific journals are routinely translated into English (with a lag of about a year), they are seldom read by scientists outside the Soviet Union. This fact was brought home to me forcibly a number of years ago when I had occasion to look at references in papers appearing in major scientific publications. Western authors almost never quoted Soviet work. Perhaps more surprising, Soviet authors (all men of international stature) quoted Soviet journals only about 25 percent of the time, the rest of the references being to the same English-language journals cited by their European and American colleagues. There could be many reasons for this situation, but the fact remains that publications by obscure scientists in Soviet journals are unlikely to attract attention in the world scientific community.

Polywater thus languished in obscurity until it was taken up by Boris Deryagin, an internationally known chemist at the prestigious Institute of Physical Chemistry in Moscow. Deryagin was so intrigued by the idea of a new kind of water that he launched a major research effort at his Institute—an effort that was eventually to involve dozens of scientists. More important, he is one of that small handful of Soviet scientists whom the authorities allow to travel to the West. Most Western scientists learned about the existence of polywater from listening to Deryagin's presentations of his work at international meetings between 1966 and 1969. Once polywater was introduced into the West, the events that followed were almost a parody of the way we handle scientific controversies in our public discourse.

The first accounts of the new discovery were straightforward descriptions of the polywater phenomenon in magazines devoted to reporting on scientific topics, such as *Chemical Engineering News* and *Scientific American*. *The New York Times*

*The name comes from polymerized water.

also ran a short article, and, on the whole, the public was about as well informed about developments on this front as it is on most basic research. There were a few tut-tutting comments about allowing the Russians to dominate this new field (a poly-water gap?), and the usual list of benefits (from improved Band-Aids to major new discoveries in medicine) that were supposed to follow an unraveling of this new mystery. All in all, there was hardly enough to cause a raised eyebrow in that segment of the public that follows science news.

All this changed in the early fall of 1969, when the prestigious British journal *Nature* published a letter from F. J. Donahoe of Wilkes College. Donahoe warned that if polywater ever got loose in the environment, it would catalyze the transformation of all the earth's water into a new form, turning the earth "into a reasonable facsimile of Venus." He argued the polywater was "the most dangerous material on earth," and urged scientists to take the greatest possible precautions in handling it. Even in those days when the environmental movement was just start-ing up, this warning caused a sensation in the press. Had this sort of thing occurred today, various groups in the United States would doubtless have gone to court to block polywater research, as some misguided individuals have done recently to block re-search in genetic engineering.

In early 1970 a regular polywater bandwagon developed. The Russians continued their painstaking measurements of the prop-erties of polywater, and American and British labs began churn-ing out hurried studies on the subject. Some claimed to find marked differences in structure between polywater and regular water (suggested structures are shown in figure 8–5 below). Theorists started to work out the molecular structure of the new form of water. Journalists headlined the threat to life on earth

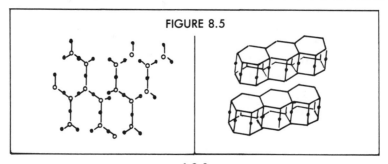

FIGURE 8.5

that was coming out of the laboratories—all this before anyone had verified that a new form of water actually existed!

The brief polywater boom dwindled as "second thought" experimental results began to be reported. The same laboratory staff at the University of Maryland that had studied the structure of polywater and declared it to be a new form of water now found that their results could be explained by the presence of small amounts of impurities in the water. They also found that if they took great pains to exclude such impurities from their samples, the evidence for polywater disappeared. Some groups claimed that polywater was ordinary water contaminated by small amounts of sweat, others that the impurities were minute amounts of silicon leached from the glass tubes. Polywater enthusiasts countered by saying that the samples in other labs may have been contaminated through improper procedures, but that their own samples still yielded the same data. On August 17, 1973, Deryagin's group finally bowed to the mounting evidence and threw in the towel. They reported that they, too, were unable to produce polywater in systems free from contamination. The world had been saved from another Doomsday threat.

Since that time, philosophers of science have studied polywater as a manifestation of abnormal science and drawn weighty conclusions from the historical record. From my point of view, there is absolutely nothing abnormal about the science in the polywater episode. A difficult experiment was done and, on the basis of the results, a claim was made. Within two years of the time that claim came to the attention of the scientific world at large, further work showed that the claim was wrong. This is as good a proof as you are likely to find that the scientific process, if left to itself, will eventually weed out mistakes. In fact, the only abnormal thing I can see in the entire polywater episode was the fact that a good deal of the argument took place in the press, rather than in scientific journals.

Unfortunately, the pattern established in the polywater debate has become all too common in our time. A new phenomenon is observed, and some scientists, more eager for publicity than for enlightenment, announce that the end of the world is at hand. In the early 1970s, the issue was the manmade threats to the ozone layer. Today, a decade later, atmospheric physicists tell us that all the observed changes in the ozone layer could easily be due to small fluctuations in the sun's brightness and hence have nothing whatever to do with human activity. More recently, the "nuclear winter" hit the headlines, with a group of scientists claiming that a nuclear exchange would turn the earth into a permanently frozen ball. Although this assertion hasn't been totally resolved as I write these lines, subsequent studies have shown that at the very least the group overstated their case. It now appears that a nuclear war during the winter would have little climatic effect beyond producing a five-day cold snap, and that the temperature drop associated with a summer war (perhaps 15°C) would last for a month rather than for eternity. These results of second thought calculations, unlike the original scares, have been largely ignored by the press.

In my darker moments, I advise my friends to follow Trefil's Law when reading about Doomsday warnings in the newspaper:

*Never believe any scientific predictions*
*made at a press conference.*

# NINE

# Snowballs in August

*Out of whose womb came the ice? and the*
*hoary frost of heaven, who hath generated it?*

—JOB 38:29

M Y FRIEND Nick Kosorok, who is responsible for keeping the road over the Beartooth Mountains passable, tells me that there is one place on the highway that is sure to be crowded any day during the summer. It's a shaded spot on the side of a slope where a large snowbank lies right next to the road. Every out-of-state car carrying vacationers to Yellowstone National Park stops there, and most people wind up taking pictures of the kids throwing snowballs in the middle of August. There is something inherently fascinating about the presence of snow where no snow ought to be.

Innocent as that snowbank looks today, it was only about 25,000 years ago that a snowbank in that same spot began to grow. A bit more snow survived each year until the accumulated mass became a huge glacier that carved out what is now Rock Creek Valley (see the photograph on p. 134). About 18,000 years ago, when the glacier reached its fullest extent, great ice

The author's daughters and father-in-law enjoy a secluded
snow bank in August. Shosone National Forest, Wyoming.

sheets covered most of North America and Eurasia, extending
well into the northern tier of states in the United States. Then,
after spreading out, the glaciers retreated, leaving only the occasional year-round snowbank and mountain glacier at high
altitudes. Given the extent of the advance of the glaciers and
the evidence they left behind in the mountains, it's a little surprising that it was only in the middle of the nineteenth century
that geologists accepted the fact that there had been Ice Ages
in the past.

Ordinary people living in the mountains knew that the glaciers
had been much larger in the past than they are now, but this
knowledge was ignored by scientists. The great Swiss-American
geologist Louis Agassiz gave the first modern statement of the
"glacial theory" in 1837, having been converted to the idea by
an old friend, Jean de Charpentier, a mining engineer. The
technique of persuasion that Charpentier used was simple, and
it's basically the same technique I will use to persuade you that
glaciers have played a major role in the development of the
earth's surface. Charpentier simply invited Agassiz to walk with
him in the Alps and showed him the evidence for the belief in

past glaciers. Agassiz was an open-minded man and, confronted with the evidence, accepted the theory. As a young geologist who had already acquired an international reputation, he was ideally suited to present the theory to his fellow scientists.

One charming incident related by Charpentier occurred when he was walking through a Swiss valley. He met an old man, a woodcutter, and the two walked along and chatted for a while. Seeing a large boulder at the side of the road, Charpentier started to examine it. The woodcutter told him that the stone came from a spot several miles away, and had been carried to its present resting place by the Grimsel glacier. He remarked casually that the glacier had once extended all the way down into the valley, almost to the location of the present Swiss capital of Bern. Charpentier commented that "this good old man would never have dreamed that I was carrying in my pocket a manuscript in favor of his hypothesis. I gave him some money to drink to the memory of the ancient Grimsel glacier."

The direct evidence for the idea that glaciers once covered our mountains is of two kinds: (i) material carried by the glaciers and deposited when they melted; and (ii) the marks left by the glaciers on rocks. The most striking evidence of the first kind is the sort of thing that Charpentier and the woodcutter encountered—large rocks transported many miles by the glaciers. Such rocks are called "occasional boulders" by geologists. As you can see from the photographs on pages 136, they can be quite spectacular—huge rocks in the middle of an open meadow or on a valley floor. Before the nineteenth century, the existence of such rocks was a complete puzzle. How could they have been moved to their present locations? You can often tell at a glance that they are identical to rocks found in large formations many miles away. The boulders obviously had been moved, but how?

A favorite explanation was that they had been moved by the great biblical flood. The hitch in this explanation was that no known currents of water could move such large objects over so many miles. A variant of the flood theory, popular for a while, was that the boulders had once been inside large floating icebergs. When these icebergs melted during the flood, the boulders were dropped wherever the waters had carried them. While this theory could explain the movement of the boulders, it had

An occasional boulder in the middle of a field. It could have arrived at this spot only through the action of a glacier. Absarokee Wilderness Area, Montana.

Another occasional boulder, this one in a valley where trees have subsequently grown. Beartooth Mountains, Montana.

A typical U-shaped glacial valley. Rock Creek, Montana.

trouble explaining why they should occur so frequently in mountain valleys and so seldom elsewhere.

The marks left by glaciers on bedrock played an important role in convincing Agassiz's contemporaries of the truth of the glacial theory. One of the most obvious of these traces is the shape of valleys that have been dug out by glaciers. When a stream cuts through a mountain, the resulting valley is V-shaped, the stream carving a relatively narrow path at the bottom. A glacier, on the other hand, gouges out a U-shaped valley, removing substantial amounts of material all across the valley floor. This shape is shown clearly in the photograph on page 137. The existence of these U-shaped valleys below the present site of glaciers in the Alps was strong evidence that the glaciers now confined to the high peaks had once worked their way down into the valleys.

As a glacier grows and starts to move, two types of processes leave marks on the rocks along its path. On the valley sides, the movement of the ice smoothes off the rocks against which it rubs, giving them a characteristic polished appearance. Polished rocks are shown in the photograph on page 138. They have a shiny surface, and once you know what to look for, you

Polished rocks line the walls of a glacial valley. Note the peaks in the background, with the polishing extending only part of the way to the top. Custer National Forest, Montana.

This jumbled pile of rocks at 9,000 feet marks the farthest advance of a small glacier. Absarokee Wilderness Area, Montana.

can't miss them. You can tell how thick the glacier was by noting how high the polishing extends. In the photograph, for example, the jagged peaks in the background probably stuck out like islands above the surface of the glacier, while the polished rocks just below the top show that the peaks only narrowly escaped being engulfed and broken off by the moving ice.

In other spots, there are deep grooves in bedrock over which a glacier has moved. These grooves are caused by large rocks that the ice has taken up and dragged across their stationary counterparts. In effect, the ensemble of glacier-plus-rock acts like a huge file, scoring the bedrock over which it moves. These grooves were well known throughout Europe, and the glacial theory provided a simple explanation of their origin.

The geological formations left behind by the glaciers when they started to melt were the final bit of evidence. As we shall see later, a glacier is not a static block of ice, but a dynamic, flowing system. The ice within a glacier moves, much like the water in a river. Rocks picked up by the ice in the "upstream" region are carried down to the leading edge of the glacier. At that point the ice melts or evaporates and the rocks are deposited. So when the glacier has reached its farthest extent, we expect to find a large deposit of material left behind both by the process just outlined and by additional material that is pushed by the front edge of the ice sheet. These deposits are known as glacial moraines; a picture of a small moraine is shown on page 138. Most moraines are larger than this, and look like small hills. In fact, the detailed maps of the various glacial advances that you sometimes see were made by discovering where the moraines are now, and then connecting them together to determine the extent of the ice sheet.

The evidence offered by moraines was less convincing in the nineteenth-century debate than you might think. The reason was that there was not one, but several Ice Ages. Each glacier has advanced and retreated many times over the past few hundred thousand years, and it is not hard to see that the advance of a new glacier would largely obliterate the deposits left by its predecessor. Consequently, the investigation and dating of moraines is at best a tricky business—not at all the sort of smoking pistol one needs to establish a new scientific idea.

If Agassiz's colleagues were reluctant to accept the glacier hypothesis when it was first presented, it was in part because few of them had trekked into the Alps to see things for themselves. But evidence accumulated, and by the 1860s when he came to the United States and joined the faculty at Harvard, Agassiz could take some satisfaction from the fact that most of his fellow geologists had come around to accepting his ideas. As happened with continental drift, once the evidence was clear, the scientific community accepted a revolutionary idea with alacrity.

## From Snowbank to Glacier

The process by which a pile of snow becomes a full-fledged glacier is a rather interesting one. The thing starts, of course, with falling snow. Anyone would suppose that the snow that falls during winter on any given spot will have melted before the first snowfall of the next winter. It is only in a few places of high altitude, such as the one shown in the photograph on page 134, that a year's snowfall survives through the summer. These spots are the breeding ground of glaciers.

If the temperatures during the summer are cool enough to allow snow to accumulate, changes occur in the nature of the snow on the ground. While it falls, snow is light and fluffy. The flakes are the familiar six-sided, lacy shapes we have all seen in photographs and illustrations. Because of the open structure of the flakes, "wild snow" (as geologists call it) has a very low density. It's hard to pack many flakes into a given volume without changing this structure, so falling snow is a fluffy, powdery substance. But almost as soon as it falls, it starts to pack and become more dense. This process occurs by the breakup of the flakes into smaller particles that have a more rounded shape. If the snow is blown by the wind, the projecting arms of the snowflakes will break off and the rounding becomes easier.

Over time, the rounded snowflakes pack down together by the force of gravity and by the weight of new snow above them. The net result is a system as shown on the left center in figure 9–1. The rounded snow crystals are packed together, but there are many air passages between them. The density of such a packing is roughly half that of glacial ice.

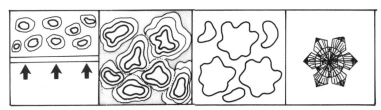

FIGURE 9.1

The next step in the evolution of the snowflake is shown in the right center of figure 9–1. Water molecules evaporate off the surface of the spherical crystals, float through the air, and then condense and freeze on open surfaces elsewhere in the pack. The result is that the air passages originally formed in the ice start to become sealed off. In this stage, the snow is called "firn," and we are almost to the point of having a glacier.

The final stage is shown on the far right. Under the pressure of the overlying snow, the firn starts to recrystallize. Much of the air is squeezed out and only occasional air bubbles remain. The density increases as the air leaves, and the material becomes true glacial ice. The only change from this point on is a compression of the air bubbles and a slight increase in density to about 90 percent that of water, the normal density of ice. A glacier has been born.

Once a glacier is formed, it starts to flow downhill slowly. We can understand the balance of the glacier movement by referring to figure 9–2. At the higher elevations, the snow accumulates each year, eventually forming a glacier. The glacial

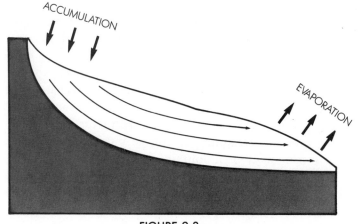

FIGURE 9.2

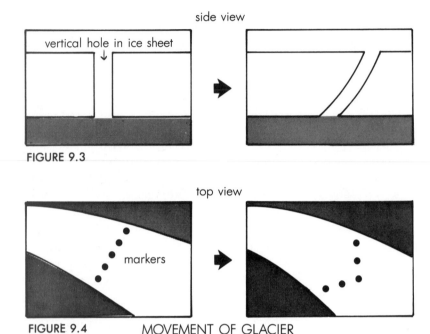

side view

vertical hole in ice sheet

FIGURE 9.3

top view

markers

FIGURE 9.4          MOVEMENT OF GLACIER

ice flows downhill until it reaches an elevation where the temperatures are high enough during the summers to melt not only the annual snowfall but the ice that is creeping down the slope. At this point, the glacier ends. The internal flow of ice in a glacier can be measured in many ways. For example, a vertical shaft can be drilled, as shown on the left in figure 9–3. After a few years, the shaft will be bent, as shown on the right, by the movement of the ice. Another way of measuring is to plant a line of markers across the top of a glacier as shown in figure 9–4. Again, after waiting a few years for the ice to move, we will find the markers arranged in a curved path as shown. Both these experiments (or other, more sophisticated versions of them) show clearly that the flow of ice in a glacier is analogous to the flow of a very thick, viscous fluid. To picture the flow of the glacier, one may imagine pouring very cold molasses at some point over a valley or mountain.

One aspect of glacier formation has given rise to an important field of research in modern geochemistry. It appears that the creation of a glacier necessarily means that air bubbles will be trapped within the ice. The air inside these bubbles is, in effect, a miniature museum, showing the chemical composition of the atmosphere as it was when the glacier formed. When we try to understand the pollution of our air by modern industry, the air bubbles in glaciers become a valuable indicator, for they show us what air was like before the industrial revolution started. If we can be sure the bubbles haven't been contaminated since they were sealed off from the atmosphere (not an easy thing to prove), then the study of bubbles in glacial ice might tell us what a "normal" environment is like. This knowledge, in turn, could be an invaluable tool for legislators and pollution control planners.

## The Ice Ages

The evidence geologists have uncovered indicates that the Pleistocene Epoch, which began about 3 million years ago and ended with the retreat of the last glaciers 10,000 years ago, had at least four major episodes of glaciation. Furthermore, there is evidence that other Ice Ages occurred before the Pleistocene, perhaps as far back as the pre-Cambrian, over 570 million years ago. The growth and retreat of the giant ice sheets, then, is not an aberration or an anomaly in the earth's history, something that happened only recently. It appears to be something that is an integral part of our planet's environment, like the alternations of day and night. This means that when we look for causes of the Ice Ages, we cannot be content with explanations that account for only one or two glacial advances, there must be some mechanism that repeats the process many times.

From our description of the way a glacier forms, it is obvious that the most important single factor affecting the onset of an Ice Age is the summer temperature in regions where snow is likely to accumulate. If the summer temperatures in the northern latitudes are low enough, snow will start to compact and form itself into ice. The ice sheets will start to spread out from many centers, some of which will be mountains. Other, larger glaciers will merely flow outward from a central source. It is glaciers of this latter type that now cover Greenland and Antarctica.

The man who discovered the mechanism by which summer temperatures in the north could rise and fall regularly was Milutin Milankovitch. Born in Serbia (now part of Yugoslavia), he is perhaps one of the most interesting and least known of the scientists who gave us our modern picture of the world. He obtained his doctorate from the Institute of Technology in Vienna in 1904 (a time when much of Central Europe was part of the Austro-Hungarian Empire). After a few years working on the design of concrete structures in Vienna, he amazed his friends by taking a job as a professor of applied mathematics at the University of Belgrade. Why a young man would choose to leave the bright lights of a cultural center like Vienna for a relative backwater may seem strange, but Milankovitch had one strong motive: the freedom to pursue his own research, the kind of freedom that can usually be found only in an academic setting. Another motive may have been the strong feelings of nationalism that were sweeping Europe and especially the Balkans at the time. In any case, while his friends shook their heads, Milankovitch packed his bags and took off for Belgrade.

There, according to his memoirs, an event occurred that changed his life. He and a friend (a poet) were in a café, celebrating the publication of a volume of the poet's patriotic verses. Apparently, newly appointed professors then weren't paid any better than they are now, because all the celebrants could afford was coffee. A man at the next table, a banker, asked what the occasion was. When he was told, he asked to see the book. As a Serbian patriot, he was so taken with the verses that he ordered ten copies on the spot. Now the friends could really celebrate! After the first bottle of wine, they "looked back on their earlier achievements, which now seemed narrow and limited." By the end of the third bottle, the poet had decided to embark on an epic. Not to be outdone, Milankovitch decided that he wanted to "grasp the entire universe and spread light to its farthest corners."

Most of us, having spent an evening like that, would have smiled ruefully the next morning and gone about our business. Milankovitch, however, decided after this experience to devote his talents to developing a theory that would explain the climates on all the planets in the solar system. He went about doing this in a methodical way, setting aside some part of every day for work on his theory, and even carrying books and papers along on family vacations. As the first wars in the Balkans started—wars that would eventually lead to World War I—Milankovitch was repeatedly called to duty as a staff officer in the Serbian army.

When the war started in earnest, Milankovitch was captured by Hungarian troops and jailed. Then something happened that would be impossible today, but which apparently caused no particular stir at the time. A member of the Hungarian Academy of Sciences arranged for Milankovitch's release, brought him to Budapest, arranged for him to have a desk at the Academy, and allowed him to continue work on his climate theory throughout the remainder of the war. It was this work, published in 1920, that brought Milankovitch to the attention of the people in Europe who were interested in explaining the Ice Ages. The most important of these were Alfred Wegener (of continental drift fame) and his father-in-law Wladimir Koppen in Hamburg. Milankovitch sent them the predictions of his theory, and they, with their access to the geological community, found that it matched the Ice Ages perfectly. The theory then developed is in essence the one we use today to explain the phenomenon.

The idea developed by Milankovitch is quite simple. We know from Newtonian astrophysics that both the rotation and the orbit of the earth undergo periodic changes under the influence of the gravitational forces exerted by other members of the solar system. The changes affect the sunlight falling on the northern hemisphere in two ways: they can change the distance between the earth and the sun, and they can change the tilt of the earth's surface relative to the incoming sunlight. Both of these effects change the amount of sunlight received at northern latitudes, and therefore can, in principle, affect the ability of snow to accumulate from one year to the next.

Thus there are three important motions that we have to consider as causes for the Ice Ages. The easiest to understand is the precession of the earth's axis of rotation. If you spin a child's top, you will sometimes notice that the top is spinning rapidly about its axis while the axis itself moves around in a lazy circle. This sort of motion is called precession. The earth is very similar to the top. It, too, spins rapidly around its axis, but the axis also moves slowly in space. Measured in relation to the sun,

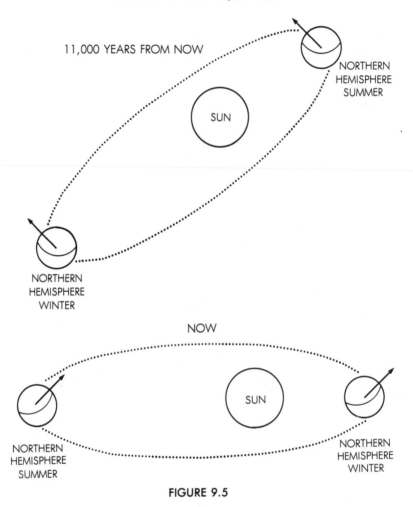

**FIGURE 9.5**

the earth's axis takes approximately 22,000 years to make one complete cycle. Today, our situation is that shown at the top in figure 9–5. When the earth is closest to the sun in January and February, the northern hemisphere is tilted away from the sun. During the summer, which occurs when the northern hemisphere is tilted toward the sun, the earth–sun distance is as large as it ever gets. This situation should result in cool summers and warm winters in the northern latitudes.

In 11,000 years, the precession will have carried the axis of rotation through half a turn of its lazy circle, and we will have the situation shown on the right. Summer will occur when the

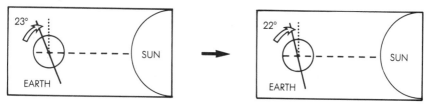

FIGURE 9.6

earth is closest to the sun and winter when the earth is farthest from the sun. Clearly, the situation 11,000 years from now will be one where summers are warmer and winters colder than they are now. If this were the only astronomical process affecting the earth's climate, we would say that the earth is currently in a period of warm, glacier-free weather.

Precession, however, is only a part of the story. Today, the axis of rotation of the earth is tilted at an angle of about 23° with respect to the plane of the solar system, as shown on the left in figure 9–6. The same gravitational forces that are responsible for the precession cause a slight rocking motion of the axis, a motion that goes through a complete cycle once every 41,000 years or so. This motion is called nutation, and the precession of the axis is superimposed on it.

Right now, the axis is moving toward the straight up-and-down direction, so that at some time in the future the earth will be oriented as shown on the right. The drawing is exaggerated for the sake of clarity—the actual change in the angle of the axis is only a degree or so. Nevertheless, it is clear that as the axis moves toward the vertical, the differences between winter and summer will tend to disappear; winters will become warmer and summers cooler. Thus at the present time the effects of nutation (cooler summers) and precession (warmer summers) tend to cancel each other out.

But the important point to note is that this is not always the case. For example, if at some time in the past the nutation was taking the earth's axis *away* from the perpendicular when the situation with the precession was as it is today, the two effects would be added to each other. Clearly, the interplay between precession and nutation can produce a varied history of warming and cooling over long periods of time.

The third important astronomical effect on the earth's climate has to do with changes in the shape of the orbit caused by the gravitational effects of the other planets. As shown in figure 9–7, the earth's orbit around the sun is not exactly circular: it is slightly elliptical. The other planets can act to flatten out the ellipse or to make it more circular. Thus, the earth's orbit goes through a cycle like that shown below about every 93,000 years. It first moves away from the elliptical to something more like a circle; then it moves back again.

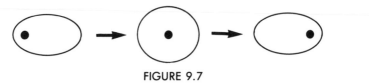

FIGURE 9.7

At the moment, the earth's orbit is becoming more circular. Obviously, differences between summer and winter temperatures would tend to even out as the distance between the earth and the sun becomes more uniform. This means that at present this particular gravitational effect is acting to cool the summers and warm the winters, just like nutation. In the past, when the movement was from circular to elliptical, differences between summer and winter would increase.

What Milankovitch realized was that by working out the effects of all three of these cyclical motions at once, he could predict how much energy in the form of sunlight had fallen on the northern hemisphere during any summer in past ages. When he did these calculations, he found a predicted curve like the one shown in figure 9–8. The deep dips in the curve, corresponding to very low summer temperatures at high latitudes, would correspond to periods when the ice sheets could grow. They also correspond very well to the estimated times of Ice Ages. In recent years, we have discovered that the motions of the earth in response to gravitational forces exerted by other

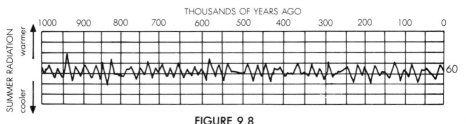

FIGURE 9.8

bodies in the solar system are the main causes of recurrent Ice Ages whose existence we can detect in the geological record.

## The Future of the Earth

One question that inevitably comes to mind in this discussion of glaciers concerns the future. When can we expect the next Ice Age? As it turns out, this is a rather difficult question to answer. There are all sorts of estimates in print, ranging anywhere from a few hundred to 20,000 years.

If we merely take the Milankovitch theory and work out when the next big drop in solar input is due, we find that it won't occur for about 20,000 years. The reason for this is the ambiguous situation we are in, with one effect (precession) predicting warming, and the other two (nutation and orbital shape) predicting cooling. Other predictions are based on shakier data: for example, by simple extrapolation of the observed cooling trend that has dominated the earth's climate since the 1940s, you find that Ice Age temperatures will be reached in a matter of 700 years. But such extrapolations on short-term data are notoriously risky. Imagine the prices we would have expected to pay today if we had simply extrapolated the double-digit inflation of the late 1970s to the present!

Besides, predictions about future climate are subject to the effect of one process that scientists are just learning to calculate—the so-called greenhouse effect. This is how it works: When fossil fuels like coal and oil are burned, carbon dioxide is produced as a byproduct. Entering the atmosphere, this gas tends to trap heat that would otherwise be radiated into space. In effect, it acts as a kind of blanket wrapped around the earth; the more carbon dioxide, the thicker the blanket and the higher the resulting terrestrial temperature. The current extremely high surface temperature on Venus is believed to be due to the accumulation of carbon dioxide in the Venusian atmosphere.

If the burning of fossil fuels continues unabated, then within a few hundred years we will have removed all the carbon atoms stored in fuels under the earth, combined them with oxygen, and put them into the atmosphere as a blanket of carbon dioxide. It is estimated that it would then take a thousand years for the atmosphere to cleanse itself of this excess material. During this time, we would expect a period of global warming.

If we are indeed at the threshold of a new Ice Age, when the Milankovitch theory would indicate a lowered temperature, it is possible that the effects of carbon dioxide might delay the inevitable for a while. In fact, so uncertain is our knowledge of the way our climate works that we cannot tell whether the near-term temperature trend ought to be upward (if the greenhouse effect wins); downward (if the Milankovitch effects win); or if the two will cancel each other out and leave everything pretty much as it is.

Strange as it may seem, there is something in the glaciers now sitting in our mountains that may shed some light on this question. We noticed earlier that the air bubbles in glacial ice contain samples of what the atmosphere was like when they were formed. An analysis of the bubbles ought to tell us what the concentration of carbon dioxide in the atmosphere was like before the great burning of fossil fuels started in the eighteenth century. This knowledge would help settle one of the difficulties encountered by people who want to predict warming due to the buildup of carbon dioxide: reliable data on the amount of the gas present in the atmosphere has been available only since the 1950s. There has not been enough time to discern long-term trends in climate. If we could tap the glacier bubbles and get data enabling us to compare amounts of the gas in that air over the last few centuries, we could see the trends more easily and, in principle at least, predict the future course of the climate. So it seems that the glaciers themselves may be able to tell us when they will start to grow again.

# TEN

# The Mountaintop Experience

*Libushje climbed to the top of (the mountain)
and prophesied: "I see a great city, whose
glory shall reach to the stars...."*

—Cosmas of Prague (1045–1125)

ALMOST EVERY cultural tradition includes the story of a hero or heroine climbing to the mountaintop to receive enlightenment. Moses receiving the Ten Commandments is one example, the prophecy of the future of Prague from Czech folklore quoted above is another. The scholar-novelist John Barth has argued that one of the common characteristics of the heroic tale in cultures from Africa, Asia, Europe, and the Americas is the journey to the mountaintop. Throughout human history, the mountain peak seems to have been identified with the experience that gives us insight into the workings of the world and into our place within it.

In a very real sense, this tradition continues today. Most of our knowledge of the physical makeup of the universe has been won at a handful of mountaintop telescope installations around the world. If we want to learn about the world, we still have to climb a mountain, as men and women have been doing for millennia. There is a new aspect to this process today, however, because for the first time it seems that we may have the capability of going well beyond the view from the top: we can now pluck our knowledge from the depths of space itself.

The best place to start this story is to recall why it is that telescopes are located on mountaintops. The purpose of a telescope, of course, is to gather light that has been emitted by distant sources. By analyzing this light, we learn about the emitting object, whether that object is a nearby ship, a neighboring planet, or a galaxy at the edge of the observable universe. Sometimes this knowledge is rather straightforward—we learn the size and shape of the thing being studied, more or less in the same way as we would learn about the things by looking at them with our naked eye. In other cases, we subject the incoming light to a more sophisticated analysis and determine facts about the objects such as temperature and chemical composition. But no matter what type of analysis we do, we are always limited by the amount of light available to our instrument.

We normally think of the earth's atmosphere as being completely transparent, for we can often see things a surprising distance away. For example, I can remember standing on a mesa near Los Alamos, New Mexico, at night and seeing the street lights of Alberquerque, some 90 miles away. Anyone who has flown across the country on a clear day has seen cities or mountain peaks a similar distance away. From such everyday experiences, we conclude that light can easily be transmitted through a thickness of atmosphere ranging up to 100 miles.

We also know that the earth's atmosphere forms a fairly thin layer around our planet. For example, when your aircraft is flying at 30,000 feet (about six miles), the cabin has to be pressurized so that you can breathe. "Outer space" is generally reckoned to start at an altitude around 50,000 feet—roughly ten miles. This means that light entering the atmosphere has to travel at most ten miles to get to the surface. Why, then, do we go to the trouble of building roads to the tops of mountains to put our telescopes a few miles closer to the source of light? Wouldn't a telescope function just as well at sea level?

The answer to this question depends on what we want to study. If we want to look at a relatively bright, nearby object, then sea-level telescopes will work perfectly well. Galileo saw spots on the sun and mountains on the moon from his balcony in Florence, and the planets Neptune and Uranus were discovered by telescopes located on the ground near major European cities (Berlin and London, respectively). Even today, the study of the positions of nearby stars (a field known as astrometry) is carried out with such telescopes, one of which is located in downtown Pittsburgh. The accurate determination of the position of nearby stars has great practical importance, since it is by sightings of such stars that satellites are navigated. Nevertheless, such studies do not occupy the main attention of professional astronomers.

A much more likely target for a telescope being used in frontier research is a distant faint object, be it star, galaxy, or interstellar cloud. What is more, since the mid-nineteenth century astronomers have not been content with finding out where objects are in the sky: they have moved on to asking what the objects are. Answering this sort of question requires a sophisticated analysis of incoming light. In general, the fainter the object is, the harder it is to collect enough light to perform the analysis. The astronomer often finds himself in a position analogous to that of a pollster trying to predict the outcome of an election by interviewing two voters—neither has enough data.

Finally, if an object is very distant from the earth, it is likely to appear to be quite small, regardless of its true size. This means that it will be difficult to see even under the best possible circumstances, and the atmosphere of cities does not provide such conditions. The problem of faintness is aggravated if we want to resolve the internal structure of the object being studied. These difficulties of faintness, atmospheric distortion, and apparent small size are what have driven astronomers to the mountaintops in their quest for knowledge of the universe.

Nor is this all. Imagine yourself on a deserted highway on a hot day, trying to make out the wording on a distant sign. There are a number of things that could make this task difficult. If the highway has been warmed by the sun, there may be rising air currents causing objects to appear to shimmer in the distance. This shimmering will make it hard to read the sign. Another difficulty will arise if the sun happens to be rising or setting near the sign. You will then have to shield your eyes from the sun, which interferes with your ability to read the sign. Both of these effects—shimmering and interference from other light sources—place important limitations on our ability to measure the stars with our telescopes. They therefore play a significant role in the decision to locate these instruments on mountaintops.

The shimmering of the horizon on a hot day and the twinkling of stars at night both arise from the same process. When air is heated, it expands like any other substance. Heated layers near the highway cause light rays to bend, as shown in figure 10–1. This effect is responsible for the common mirage you

FIGURE 10.1

see when driving: the road ahead of you seems to be covered with water, though when you arrive at the "puddle," the surface is perfectly dry. In that case, the light ray is bent as shown, and what you are actually seeing is the open sky. The sky appears to be located on the road surface because the deflected light beam enters your eye at an angle and you interpret this as a light signal coming from the road itself.

If, instead of the relatively stable layers of heated air shown in figure 10–1, the sun's warmth causes a series of rising air bubbles (see chapter 1), we get a situation like that shown in figures 10–2a and b. As a bubble rises, the rays of light from the objects on the horizon are momentarily deflected. When this happens, as shown in figure 10–2a, the light from the horizon appears to be coming from a point below its actual origin. A moment later, when the bubble has passed, we have the normal situation shown in figure 10–2b. The passing bubble, then, results in an apparent, albeit temporary, shift in the position of a distant object. It is the bending of light rays in rising air bubbles that causes the shimmering of the horizon.

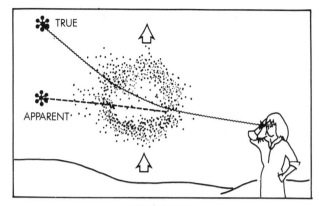

**FIGURE 10.2a**

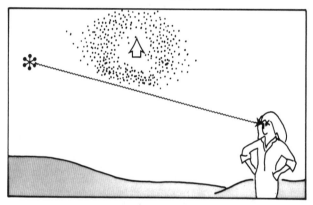

**FIGURE 10.2b**

Now let's turn to the study of the stars at night. It may be surprising to think of the cool night air in the same way that we think about air over the hot concrete of a highway, but the difference is a matter of degree, not of kind. After sunset the air cools quickly, but the ground remains relatively warm. This disparity results in small bubbles of heated air rising through the atmosphere. The effect is the same as it was for the highway; each time an air bubble passes our line of sight to a star, the position of the star appears to change slightly, as shown on the left in figure 10–3. The result (as seen by the observer) is shown on the right. The position of the star keeps jumping around. Our eye, marvelous instrument though it is, cannot keep track of the rapid fluctuations, and what we actually see is a blurred-out, twinkling point in the sky.

APPARENT POSITION
OF STAR

INTEGRATED
IMAGE

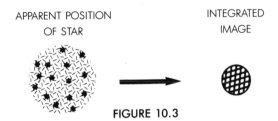

FIGURE 10.3

At sea level, the light has to travel through many miles of atmosphere, encountering many temperature changes on the way. An observer on the ground sees the light shift over ten times a second, so no clear image is possible. One advantage of a mountaintop observatory is obvious—there is simply less air between the telescope and the object in space. Hence the images of stars are much sharper than those same images seen from the ground. This statement can be verified by anyone who has ever spent a night under the mountain sky. This advantage is enhanced by the fact that the density of the atmosphere decreases with altitude, so that high mountain peaks are like islands sticking above the densest parts of the atmosphere. Even though the atmosphere extends to a height of 50,000 feet, in the upper regions the molecules are few and far between. By the time you have reached 15,000 feet, all but a small part of the atmospheric mass lies below you, and the less air there is, the less bending of the light there will be.

Another advantage of mountaintops is that they tend to be far removed from extraneous light—lights that might play the role of the setting sun in our highway example. While it isn't always true that mountaintops are free from light pollution (a point to which we'll return later), it is certainly possible to find mountains far from centers of population. This matters, because our ability to distinguish faint objects is limited by the extraneous background light in the field of vision. For example, when you are in an airplane flying over the countryside at night, you can often pick out a light in an individual farmyard on the ground, even from altitudes of 10,000 to 20,000 feet. You can do this because even though the light itself is relatively feeble, it does not compete with any background illumination. If you were flying over the same spot during the day, you couldn't even tell whether there was a light on or not. The daylight would wash out the feeble signal from the farmyard. In exactly that way, the light from faint objects in the sky is easily lost in the glow of city lights or the light of the full moon.

In a very real sense, the beginning of twentieth-century astronomy can be traced to the installation of a 100-inch telescope atop Mount Wilson near Los Angeles in the early 1920s. Edwin Hubble, using the unprecedented resolving power of his new instrument, was able to establish that our Milky Way was only one of a large number of galaxies in the universe. The technique he used was simple, at least in principle.

A certain class of stars known as Cepheid variables grow alternately bright and dim over periods of weeks or months. By examining such stars in the neighborhood of the earth, it had been established that the time it took each star to go through one if its cycles depended on how much total light it emitted. This means that if you measure the amount from the variable star light that falls on the telescope, you can tell how far away it is. If the star appears faint while its period indicates that it is actually very bright, you conclude that it must be far away. Similarly, if it appears bright but its period indicates that it must be faint, it must be close by.

Because Hubble could actually pick out individual stars in the Andromeda Nebula, he was able to establish that its distance from the earth was over 200 million light-years. Contrasting this 200 million to the greatest distance between stars in the Milky Way— a "mere" 100,000 light-years—Hubble argued that the collection of stars in the Andromeda Nebula constituted an "island universe" like our own. Overnight, the horizons of astronomy expanded from the study of a single galaxy to the study of the collection of billions of galaxies that make up our universe.*

Hubble's discovery, with its immense implications for the place of mankind in the universe, depended on his ability to make out a single star in a galaxy hundreds of millions of light-years away. This is a familiar pattern in the history of science; the greatest intellectual advances often depend on our ability to deal with the grubby details of instrumentation. In Hubble's case,

*This same study also showed that the galaxies in the universe were moving away from each other, a result that gave rise to our picture of an expanding universe originating in a Big Bang. You can read more on this topic in my book *The Moment of Creation*.

success depended on his ability to distinguish a very distant, very faint object. To do this, he needed both a state-of-the-art telescope and a mountaintop on which to put it, well above a good portion of the atmosphere.

Unfortunately, both the 100-inch telescope on Mount Wilson and the famous 200-inch instrument on Mount Palomar—two of the most productive instruments in the history of astronomy—have become seriously compromised by lights from the cities of Los Angeles and San Diego. Those cities weren't very large when the telescopes were built, but over the last half century the population of southern California has mushroomed, and with it the amount of illumination needed to brighten the evening hours. Scientists call this the problem of "light pollution," and regard it as the *bête noire* of modern astronomy. Fortunately, San Diego seems disposed to help keep the telescopes functional by regulating the amount and types of lighting it will permit. It is in the nature of things, though, that these two telescopes are bound to be engulfed at last by the sea of humanity swelling around them.

Partly as a consequence of this fact, the centers for observational astronomy in the United States have moved to the deserts of southern Arizona and to a high mountain peak in Hawaii. The southwestern desert has many advantages, even though the peaks there are not particularly high. For one thing, until quite recently the population of the region was small. For another, the weather in the area tends to be clear and free of clouds, a fact that affords these telescopes more viewing time than most.

The peaks on which the telescopes are located tend to be rather more cluttered than you might expect (see the photograph on p. 162). The reason for this is simple: it is expensive to develop a remote site for use as an observatory. A road has to be built and maintained, often through rugged and inhospitable terrain. Utilities, water, and housing have to be installed at the top. This concentration of facilities on a few peaks makes it doubly difficult for astronomers to deal with social trends like the growth of retirement communities in the desert. Despite the best efforts of local governments in the area, an appreciable fraction of the sky in southern Arizona has already been lost to light pollution. In response to this situation, planners are already starting to survey peaks farther to the east, in regions that are as yet unsettled. It's clearly only a matter of time before all the best telescope sites in the United States will be compromised by the spread of cities.

From this discussion, it's obvious that there are two major problems confronting astronomers in their eyries. The first is that to study the heavens, they have to look through layers of distorting atmosphere. This limitation is fundamental; it is imposed by the laws of nature. The second problem, light pollution, is not fundamental in the same sense but is nonetheless just as damaging and just as unavoidable as the first.

There is a third problem with telescopes located at the bottom of the atmosphere, one that we haven't touched on so far. We have mentioned that the atmosphere is transparent to ordinary light. What we haven't discussed is that this is something of an anomaly, for the atmosphere is opaque to almost all other kinds of radiation. This is of great significance, because most objects we wish to study emit not only visible light but many other kinds of radiation as well. The fact that only some of this radiation reaches the earth, or even the tops of mountains, means that so long as our astronomy is planet-based, certain information will be forever denied to us.

To understand what these other kinds of radiation are and why they are important, think of what you see when you watch a coal in a fire. As the fire heats up, the coal first glows dull red, then orange, then white. If you are watching the coals in a blacksmith shop, where there is a large bellows to force air through the fire, you might occasionally see a particularly hot coal take on a slightly bluish tinge. These observations tell us that the sort of light emitted by an object depends on that object's temperature. At low temperatures, that radiation tends to be reddish light, while at high temperatures it tends toward the blue. The only difference between the red and the blue light is the wavelength of the light wave itself. The red wave is about 8000 atoms long, the blue is about 4000. Red and blue light, in other words, are no more different than two different sizes of waves in water.

But the coals in your fire don't emit only visible light. When the fire has gone out and the coals have stopped glowing, you can put your hand to the fire and feel warmth. What is happening at that point is that the coals are emitting radiation with a wavelength even longer than that of red light. We call this sort of thing infrared radiation, and, while our eyes are not able to see it, we can detect it with our hand or with specialized instruments. Infrared radiation is one of a large class of waves that make up what we term the "electromagnetic spectrum." All members of this class are waves like light, but have longer or shorter wavelengths.

If the coal in your fireplace were allowed to cool further, the wavelength of the radiation it emitted would get longer and longer, running through microwave to radio. Similarly, if we allowed the coal to get very hot indeed, its radiation would soon pass out of the visible region at the other end. It would emit wavelengths shorter than blue light, which we call ultraviolet radiation. It is this radiation in sunlight that causes sunburn; so, as was true of infrared, our bodies can detect ultraviolet even though our eyes cannot. At still higher temperatures, the wavelengths would become so short that we would say the coal was emitting X-rays. All of the waves, from radio to X-rays, make up the electromagnetic spectrum. The major components of the spectrum are listed below.

| THE ELECTROMAGNETIC SPECTRUM | |
|---|---|
| TYPE OF RADIATION | APPROXIMATE WAVELENGTH |
| gamma rays | the size of an atomic nucleus |
| X-rays | the size of an atom |
| ultraviolet | hundreds of atoms |
| visible light | 4000–8000 atoms |
| infrared | tens of thousands of atoms |
| microwave | an inch |
| radio | up to thousands of miles or more |

The point of this discussion of the electromagnetic spectrum is to establish a generality: Every object in the universe emits some sort of radiation. The very hot objects emit energetic, short wavelength radiation like X-rays; "normal" stars like the sun emit waves primarily in the range of visible light; while cooler objects emit in the infrared and radio ranges. All this radiation, indicative of a wealth of information about all sorts of things in the sky, pours down continuously on top of the earth's atmosphere. What secrets we could learn if we could interpret it!

Unfortunately, this isn't easy. As we intimated earlier, the atmosphere absorbs most of the electromagnetic spectrum. There is a "window" of transparency that allows visible light to travel long distances in the air, another window that allows some radio waves to do the same—and that's about it. Every other type of radiation from ultraviolet to infrared is absorbed by the atmosphere long before it can reach the ground. Any telescope on the earth, whether on a mountaintop or in a city basement, is forced to look at the sky through an atmospheric blanket that absorbs everything but visible light and radio waves. This is why, until quite recently, astronomy was done either with large optical telescopes of the type we've been describing or with huge, dish-shaped radio receivers.

Having said this, I must add that there are a few exceptions to such a generally unsatisfactory state of affairs. Infrared radiation is absorbed by water vapor in the atmosphere, so it can't penetrate from space to sea level. Anyone who has ever climbed a mountain, particularly on the humid east coast, knows that as you climb, the humidity drops rapidly. Molecules of water tend to cluster at low altitudes. Consequently, observatories on mountain peaks tend to be above most of the atmospheric water vapor, and a good deal of the infrared radiation that doesn't make it to sea level does reach places like Kitt Peak (see the photograph on p. 162) or Mauna Kea in Hawaii (where the altitude of the observatory is about 14,000 feet). With this exception, the general rule holds: the atmosphere blocks out all radiation coming to us from the universe except for visible light and radio waves.

In spite of this rule, the past few decades have been a golden age for astronomy, perhaps the greatest that science has ever known. How so? Starting in the 1950s with primitive rockets capable of lifting instrument packages above the atmosphere for a few minutes at a time, our ability to place instruments above the absorbing blanket that surrounds us has grown steadily. The result is that one by one we have been able to open

Telescopes dot the developed site of Kitt Peak National Observatory near Tucson, Arizona. The peak in the background is called Baboquivari (the needle that connects earth and sky) and, according to the folklore of the Papago Indians, marks the center of the universe.

the windows in the electromagnetic spectrum and see the heavens in their full glory.

We have found that there are many sources of X-rays and ultraviolet light in the sky. These regions mark spots where temperatures are unusually high and where physical processes are unusually violent. Some of these regions are in our own galaxy—places where we believe that stars are being formed. The X-ray sources are not always easy to interpret, but many astronomers believe that at least one of them consists, in part, of a black hole.

Black holes are objects so compact and so massive that not even light can escape the gravitational pull at the surface. If the sun were a black hole, it would be only a mile to two across. A black hole can be thought of as a kind of cosmic one-way street: matter and radiation can fall into it, but once inside, they can never return to the universe. A black hole can never be "seen" in the conventional sense, because any light that comes near it is instantly absorbed (hence the use of the term "black"). Our theories suggest, however, that when charged particles like electrons come near a black hole and start to fall in, they will radiate energy in the form of X-rays. There are a few X-ray sources in the sky that fit the expected patterns, and astronomers think that they may be double star systems in which one of the partners is a black hole. The idea is that the intense gravitational forces exerted by the black hole are pulling material from the normal star, and as this material falls in, we detect its radiation. Thus, the X-ray part of the spectrum gives us evidence—circumstantial though it is—that black holes really do exist in nature.

For technical reasons, the exploration of the X-ray spectrum is more advanced than that of other types of radiation. The first X-ray probes were carried up by rockets in the early 1960s. They managed to peek at the sky for periods of up to five minutes, but no more. Later, the Uhuru satellite, launched in 1970, performed a rough survey of the X-ray sky, and the Einstein satellite, launched in 1978 (it ceased transmission in 1982), provided us with a more detailed map of the regions in our universe that emit in this part of the spectrum. A permanent orbital X-ray observatory is being designed right now. Called AXAF (Advanced X-ray Astronomy Facility), it will probably be completed and launched in the early 1990s.

Two satellites, Copernicus (1972–81) and the International Ultraviolet Explorer (launched 1978), have done a survey of the ultraviolet regions of the sky. In sum, then, the short wavelength, high-energy ("hot") end of the electromagnetic spectrum has already been explored pretty extensively.

For the longer wavelength parts, particularly the infrared, the state of the art is not nearly so well advanced. The reason for this is not hard to find. As we have seen, cool objects emit radiation in the infrared. If we put a detector inside a telescope, the telescope itself will radiate energy into the detector and there will be no way to distinguish between infrared signals coming from the stars and the same signals coming from the telescope itself. The only way to keep the instrument from "seeing itself" is to cool everything down to the point where the emitted radiation is beyond the infrared range. This can be achieved by enclosing the entire apparatus in a bath of liquid helium, which has a temperature of $-269°$ C, only $4°$ above absolute zero. If this is done, an infrared telescope in space can indeed be made to work.

But this happy solution at once generates a new set of problems. Liquid helium, no matter how well enclosed and insulated, will eventually evaporate. Without some way to replenish the supply, the telescope will stop working when the helium runs out.

The first infrared satellite was called IRAs (InfraRed Astronomy Satellite). It was launched in January 1983, and the helium on board lasted until November of that year. During its brief lifetime the satellite showed that the infrared sky was much richer than anyone had suspected. There were cool dust clouds where no one had suspected them, clouds of debris around nearby stars that might be connected with the process of planet formation, and wispy, cirrus-like clouds which, apparently, permeate our galaxy. As with X-ray astronomy, NASA is hard at work on a permanent orbiting infrared observatory known by the acronym SIRTF (Satellite InfraRed Telescope Facility); it is scheduled to go into orbit sometime in the 1990s.

The trend, then, seems to be this: After they've had a few preliminary peeks at the universe at wavelengths normally not detectable from the ground, astronomers want to set up a permanent orbiting observatory for the detailed study of celestial objects. The first of these extraterrestrial observatories will be launched in 1986, and will orbit the earth. It is, of course, the Space Telescope (ST). Also called the Hubble Observatory, it can be thought of as the first of a long string of instruments that will free astronomy from the limits imposed by the atmosphere. In effect, the ST will allow us to look into the heavens with a resolving power ten times that available in any telescope on the earth's surface. When pointed at the outer planets, for example, it will produce photographs as clear and as rich in detail as those produced by the recent NASA space probes. With ST, astronomers will be able to examine the detailed structure of those distant objects, like quasars, that they believe hold the clues to the secret of the origin of the universe.

Having made this point about what the ST can do, it's necessary to understand what it cannot do. For one thing, it is designed to detect only visible radiation, with a small overlap into the ultraviolet. Therefore it will not explore and map new

The NASA Hubble Space Telescope will detect objects fifty times fainter than now possible. Scheduled for launch during 1986, the 25,500-pound, 13-meter (43-foot)-long space telescope will be the largest scientific instrument ever placed in space. It will operate as an international facility with an expected lifetime of fifteen years. (COURTESY PERKIN-ELMER CORPORATION)

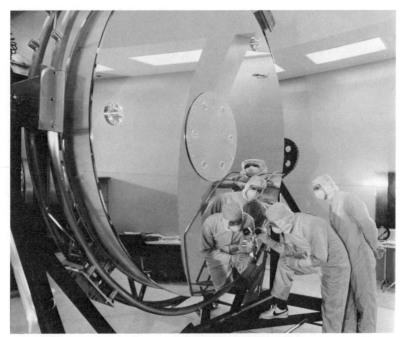

The "crown jewel" of the NASA Hubble Space Telescope, and the major element of the Optical Telescope Assembly built by Perkin-Elmer Corporation, is a 2.4-meter (94-inch) primary mirror. It is the most perfect mirror of its size ever made, and is coated with a reflective layer of pure aluminum 2.5-millionths of an inch thick, protected by a layer of magnesium fluoride 1-millionth of an inch thick. (COURTESY PERKIN-ELMER CORPORATION)

areas of the electromagnetic spectrum. For that task, we shall have to wait for AXAF and SIRTF. It is also clear that ST will not completely displace mountaintop telescopes, even in the optical region.

This last statement may seem to contradict what I've said about the limitations imposed on astronomy by the atmosphere. There is no contradiction, however. Astronomers require two things of their instruments—the ability to see objects clearly and the ability to detect very faint ones. Under the first requirement, ST will surpass any earthbound instrument, at least for a while. Whether advances in electronic and image processing will eventually allow earthbound astronomers to compensate for atmospheric effects is a matter of some debate in the community. But there is no question that even with the present technology, ground-based telescopes can see much fainter objects than will be detectable with ST. To see faint objects, we need a large light collector—what astronomers call a "light bucket." Because it has to fit inside the shuttle, the mirror on ST is only 94 inches across; this gives it only a fraction of the collecting power of a large ground-based telescope. Since objects near the edge of the universe tend to be faint, this important area of research is likely to continue to be dominated by mountaintop telescopes.

In the words of Fred Chaffee, director of the Smithsonian Institution's Multiple Mirror Telescope, "When you want to see faint, you just have to collect the light. Eventually, area wins." Mountaintop observations will doubtless be with us for some time to come.

# ELEVEN

# By a Mountain Stream

*"Time is like a river what flows endlessly
through the universe, and you couldn't step into
the same river twice..."*
*"Explain yourself, Heraclitus."*
*"You could go down to the river, and you
could step in, step out, and step in again, but
that river you stepped in would have moved
downstream, and if something were on the top
of the water, for example a water bug, it would
be downstream. Unless, of course, it was
swimming upstream, in which case it would be
older, and a different bug."*

—SEVEREN DARDEN,
*The Philosophy Lecture*

SPAWNED IN the melting icepack in the high peaks, formed from countless rivulets in alpine meadows, mountain streams come cascading down into the valleys in a shower of whitewater. So great is the force of the water that the streams play a major role in wearing down the mountains on which they form. Their rapids are important to all sorts of sportsmen, from fishermen to kayak enthusiasts. As it happens, you can also learn a lot about science by sitting at the side of a stream and watching the water.

I was doing just this one afternoon when I noticed something very strange. We are all used to the fact that when water flows swiftly over a rocky bed, there will be a lot of foam and bubbles generated—this is what is meant by "whitewater." A typical stretch of whitewater is shown in the photograph opposite. The whitewater here forms a classical wave, except for one thing. The stream in the photograph is moving to the right, and the

The stream in this photo is flowing to the right, yet the whitewater is on the left of the wave. This is an example of hydraulic jump.

turbulent crest is on the left-hand side of the wave. The wave, in other words, is pointing *in the wrong direction!*

This is a simple observation, one that you can make for yourself if there is a stream near your house. But if you think about it for a while, you'll realize that there is something strange about the waves you see in whitewater. Your instincts tell you that a rock in the stream bed should push the surface of the stream up, and if a wave forms, it should break on the downstream side of the rock. It definitely shouldn't produce foam on the upstream side.

Like most people, I had seen this sort of effect for years without really noticing it. Once I did notice it, however, I was greatly puzzled. After all, I make a point of telling my students that the laws of physics explain everything in the universe. I have no trouble explaining the behavior of distant galaxies or the inner mysteries of quarks to them; but here, by a mountain stream, was a simple phenomenon I didn't understand.

Well, if you can't understand something, the first thing to do is to observe carefully and try to find out all you can about it. I watched for a while and noticed that the wave wasn't stationary. It would appear, last for anything from a few seconds to half a minute, then disappear; it seemed to be washed downstream. Then it would re-form and go through the whole process again. I waded a little way out into the stream so I could feel the bottom with a stick. (In retrospect, I realize that this wasn't a very smart thing to do and I wouldn't recommend it to anyone else—whitewater is dangerous!) Imagine my surprise when my probings revealed that the bottom under my intermittent wave was level. The wave wasn't caused by a rock at all, but seemed to appear spontaneously above a perfectly normal part of the stream bottom. There was, however, a large rock upstream from the wave—as well as from all the other waves in the stream.

Once I had made these observations, something I'd learned in a fluid mechanics course years ago floated to the surface. The wave was an example of "hydraulic jump," a phenomenon so common in everyday life that we usually don't notice it.

Before explaining what the jump is, let me mention other places you might run into it. Next time you run your kitchen faucet, look closely at what happens when the water hits the flat bottom of the sink. If the tap isn't going full blast, you will see a circular pattern in the water. Near the point of impact there will be a thin layer of outward moving water, a layer which thickens suddenly in a circle some distance from the point of impact (see figure 11–1). You can see the same effect when

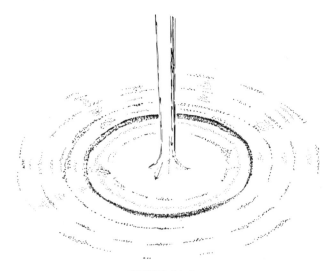

FIGURE 11.1

A hydraulic jump at the beach. On the left you can see an outgoing wave, and the straight line of foam is the jump. Point Reyes National Seashore, California.

you pour cooking oil into a flat skillet. The circles are hydraulic jumps, caused by the same mechanism as the wave in a mountain stream. Another example of a hydraulic jump can sometimes be seen at the shore. If the beach has a very gradual slope, and if the surf isn't too high, you may get a situation like the one shown in the photograph above. As one wave runs off down the beach, another starts up. The incoming wave doesn't have the normal curved shape of a breaker, but instead advances slowly in the shape of an almost vertical wall of water a few inches high. That wave is noisy, too. It has a roaring sound unlike anything else at the beach (a fact which indicates that the kinetic energy of the wave is somehow being converted into the energy of sound waves). This particular kind of hydraulic jump is known as a collapsing breaker.*

The small breaker you see in the photograph is actually a special case of a more spectacular phenomenon known as a tidal bore. In places like the Bay of Fundy in Newfoundland, the

*For a more complete discussion of the various ways that waves can come in to a beach, consult my book *A Scientist at the Seashore* (Scribners, 1985).

incoming tide encounters a large funnel-shaped inlet. In order to accommodate the flow in a narrower and narrower channel, the water becomes deeper and deeper. Eventually, you have a situation in which a wall of water 30 to 40 feet high is coming up the Bay at speeds of 20 miles per hour or more. Obviously, anyone with a boat on the Bay has to be aware of this phenomenon if he or she is to avoid disaster.

With so many examples of hydraulic jump around us, it is perhaps surprising that the phenomenon is normally discussed only in the back pages of engineering textbooks. This state of affairs has probably come about because so many strange things can happen when fluids flow that one more hardly seems important to the expert. Actually, as we shall see, hydraulic jump is analogous to something that is much more widely known— the phenomenon of sonic boom. So, if we are going to understand what we see in a mountain stream, we're going to have to dust off the old books and understand hydraulic jump in all its complexity.

First, we have to understand two simple facts about water or any other fluid. One is that at any given speed the amount of fluid being carried along is greater for a deep channel than for a shallow one. The other is that waves can be generated on the surface of a moving fluid, and the speed of these waves will depend on the kind of motion the fluid is undergoing.

The first point is easy to explain. If you want to move a group of people down a hallway, you can get twice as many through if they move in two rows instead of one. In the same way, a foot-deep stream moving at 10 miles per hour will move only half as much water as a two-foot-deep stream moving at the same speed. In general, the amount of water that has to be moved through a particular stream or riverbed is fixed by the amount of snow that melts each day or, in the plains, by the amount of water discharged from springs in the area where the river rises.

Given the fact that the total amount of water moving through a specific part of the stream is constant, it follows that the depth of water in the bed will adjust itself automatically. If the water is moving quickly (for example, if it is on a steep slope), the depth can be less than if the water is moving slowly. So long

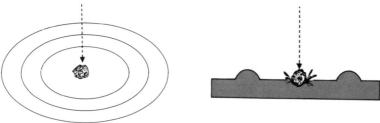

FIGURE 11.2

as the same total amount of water moves through the system, fast-and-shallow works as well as slow-and-deep.

The existence of waves on water is a familiar phenomenon. If the surface of a still pond is disturbed, we know that a wave propagates outward from the point of disturbance, as shown on the left in figure 11–2. If we took a slice through the water, that outgoing wave would have the appearance shown on the right: one wave profile moving to the left and another moving to the right.

For our purposes, the most important thing about the existence of waves is the fact that the speed of a wave on water does not depend on the way the wave is generated. Whether we drop a stone, strike the water with our hand, or allow a bubble to rise to the surface, the result will be the same outgoing circular wave shown in the figure. Once a wave is created, it leads a life of its own. Its velocity depends on the properties of the fluid in which it moves, not on the condition of its birth. In a stream, where the waves are typically several feet long and the depth in a whitewater region is a foot or less, the speed of the waves depends only on the depth of the water—the deeper the water, the faster the wave.*

If we are going to understand what we see in a mountain stream, we shall have to talk about what happens when the water on which the wave travels is itself moving. Actually, this doesn't add much of a complication. All we have to do is to recognize that if the water is moving, the circular pattern that appears in a still pond will be swept downstream. The cross section of one such case is shown on the top in figure 11–3.

---

*To be precise, if the speed is measured in feet per second and the depth in feet, the velocity of the wave is given by $\sqrt{32h}$, where h is the depth.

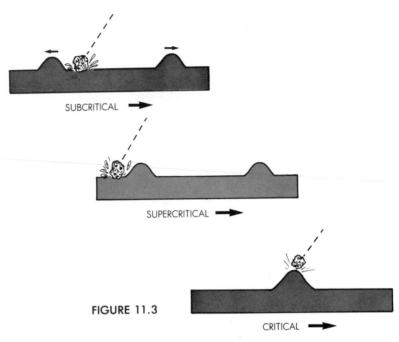

FIGURE 11.3

If the movement of the water is to the right, the downstream part of the wave will travel farther from its point of origin (with respect to someone standing on the bank) than the upstream section.

The important point is that there are two velocities involved in a wave on moving water, and these two velocities do not depend on each other. One velocity is that of the overall motion of the water downstream, a velocity determined by the amount of water than has to run through the channel. The other velocity is that of the wave, and this depends only on the depth of the water. Thus, we can have three distinct cases, each of which is shown in figure 11–3:

1. *subcritical flow*—the velocity of the water is less than the velocity of the wave, so that it is possible for a disturbance to create a wave that travels upstream against the current. This case is shown on the top.

2. *supercritical flow*—the velocity of the water is greater than the velocity of the wave, in which case nothing can travel upstream. This is shown in the center.

3. *critical flow*—the velocity of the water is equal to the velocity of the wave. In this case, the downstream wave will be swept away, but the upstream wave will be stationary as seen by someone on the shore. The upstream motion of the wave is exactly canceled by the downstream motion of the water. This situation is shown on the bottom.

Note that it is possible for the same stream to be both super and subcritical at different points along its course. If we have a shallow and fast region at one place, the flow may be supercritical there. In a slow and shallow region, the flow may be subcritical. Note also that the designation of sub- and supercritical does not depend on the absolute velocity of the stream, but on the velocity of the stream compared with that of a wave. A swiftly moving, shallow stream may be supercritical because wave velocities are low, while another stream which is moving at the same speed but is deeper may be subcritical, simply because wave velocities are higher.

With this background, we know enough about the nature of fluid motion to understand hydraulic jump. The key to what got my attention in the mountain stream was the existence of a large rock upstream of the whitewater. As shown in figure 11–4, the water coming over the top of the rock is flowing down a slope much steeper than that which describes the general geometry of the stream. Consequently, on this part of the rock the flow of water will be fast; and since the amount of water that has to pass over the rock is roughly the same as that which has to flow through any other area of the stream of equal cross section, the swift flow down the steep side means that the water flowing over the rock will be shallow compared to the rest of the stream. You can verify this conclusion for yourself by noting that you can often see through the water cascading over the downstream side of the rock, while you may not be able to see the bottom of the stream at any other point. This is possible because the water layer of the rock is thinner than elsewhere.

Once the water coming down the rock encounters the slower fluid in the main part of the stream, it slows down. You have a similar experience when you run to catch a plane, then arrive at the ticket counter where movement is slower. When this happens, the number of people at the slowdown spot grows; a

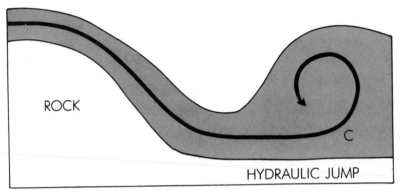

ROCK

C

HYDRAULIC JUMP

FIGURE 11.4

crowd accumulates. In just the same way, when the swiftly moving water slows down, water accumulates. In an open stream, the only way to accommodate more water is for the overall depth to increase. Thus, some distance from the rock we expect the depth of the water to be greater than it is where the water is sluicing down to the base of the rock. The only remaining question, then, is whether this change of depth takes place gradually or suddenly.

It's not hard to see that we can have a situation in which the fast, shallow flow down the rock is supercritical while the deep, slower flow downstream is subcritical. If this is the case, then there must be some spot (labeled C in figure 11–4) where the flow is exactly critical. If a wave is generated in this region, it will move upstream as fast as it is being swept downstream, thus appearing to be stationary to someone on the bank.

But why should a wave arise at all? Theoretically, if the bottom of the stream were absolutely smooth, there would be no wave. In a real stream, however, the bottom is composed of pebbles and sand. Water flowing over this rough bottom will always be generating waves. You can generate this sort of wave next time you take a bath—just move your hand up and down underneath the surface and watch the ripples in the tub.

Under ordinary circumstances, the multitude of tiny waves generated by a flow over rough bottoms cancel out in the overall motion of the stream surface. But at a point where the flow is critical, the upstream wave stands still; which means that successive waves generated by a given pebble on the bottom will stay together, reinforcing each other until the sum total is a large

*176*

stationary wave in the stream. It is this sort of large stationary wave, shown opposite, that is called a hydraulic jump. To generate such a wave, we need only have a transition between supercritical and subcritical flow. The ordinary process of wave generation by a rough bottom will provide the rest of the ingredients from which the jump is built.

Hydraulic jumps need not necessarily be large and turbulent like the one shown. Depending on the rate of flow of the stream, the jump can be anything from an almost unnoticeable ripple to a large standing wave. The hydraulic jump in the kitchen sink (which we'll explain in a moment) is almost invariably of the mild variety.

Once we understand how a hydraulic jump originates, there is no problem in understanding the observations I described earlier. There is nothing special about the bottom of the bed under the point of the jump—it's just the place where the flow passes through criticality. So we shouldn't be surprised that probing the bottom with a stick reveals no large rocks though, to be honest, I certainly *was* surprised when I first discovered this fact.

Similarly, the intermittent nature of the jump is easy to understand if we remember that a swiftly flowing stream has a rough and uneven surface. The water level on the upstream side of the rock can be expected to fluctuate erratically as troughs and valleys in the stream wash over it. If the water level behind the rock drops, the amount of water sluicing down its face will drop as well. In the most extreme case, the face of the rock will be momentarily uncovered, and the jump will disappear.

Although it may not be intuitively obvious, the jump will also disappear when the water level on the upstream side of the rock goes up. In this case, the amount of water that has to be carried down the rock increases, which means that the level of the water coming down the rock increases. Even though this thicker layer may be moving as fast as at any other time, the simple fact that it is deeper than usual means that the wave velocity in it is high. Consequently, when the water level behind the rock goes up, the flow down the rock face can become subcritical *even though the water doesn't slow down.* When this subcritical flow hits the main body in the stream, it will slow down and the overall level of the water will be increased. Yet there is no point during

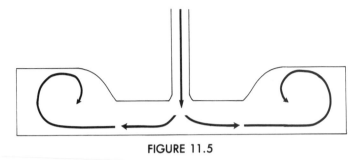

FIGURE 11.5

this process where the flow becomes critical. It simply drops from one subcritical value to another. Consequently, any waves created in the downstream region will move away from their point of origin and will not reinforce each other to build up the type of stationary wave we have called a hydraulic jump.

By this reasoning, we can see that the jump will disappear when the water level upstream increases, and again disappear when that water level decreases. It is only for a relatively narrow range of water levels, then, that the flow down the face of the rock will be exactly what it has to be to produce critical flow and a stationary jump.

The hydraulic jumps we described in the kitchen sink and frying pan can be explained in exactly the same way as the hydraulic jump in the mountain stream. All we have to do is substitute in our minds the fluid in the vertical column in figure 11–5 for the thin stream of water flowing down the rock. As the fluid in the column hits the sink or the frying pan, it flows quickly away from the point of impact. This flow is supercritical because the fluid is moving rapidly. As the friction between the sink and the fluid takes its toll, the speed drops and the flow becomes subcritical. We have the same situation we encountered before: near the point of contact the flow is supercritical, far away it is subcritical, and somewhere in between it must be critical. At the critical point, the stationary waves build up and the depth of the fluid changes abruptly. The only difference between the hydraulic jump in a stream and the jump in the kitchen is that in the latter case the fluid is flowing radially outward from a single point so that the jump is a circle, while in the stream channel the jump is a short line perpendicular to the direction of the stream flow.

At the beach, the jump is more similar to the way it is at the stream. The sloping beach takes the place of the rock; water running down the slope takes the place of water on the down-stream face of the rock; and the incoming wave takes the place of the main body of water in the stream. Once this identity is made, the mechanism for the formation of the jump is exactly the same. The tidal bore in the photograph on page 171 is highly turbulent, just like the jump in the mountain stream shown on page 169.

In principle, it ought to be possible to have tidal bores that are not turbulent—bores that closely resemble the hydraulic jumps in the kitchen and frying pan. I have never seen that sort at a beach; in fact, the bore in the photograph is the first I ever saw in my life. I have been told, however, that smooth, non-turbulent bores can be seen where the incoming waves are small and gentle—places like beaches on the Great Lakes. If you should run into them somewhere, why don't you write and let me know?

The connection between the hydraulic jump in the stream and a sonic boom is a little less direct than that between the jump and other phenomena we have been discussing. The essential point of similarity is that both depend for their existence on the reinforcement of waves generated at different times. In the hydraulic jump, these waves reinforce each other because at criticality all upstream waves are stationary. In the sonic boom, the details are slightly different, as shown in figure 11-6.

At each instant, the sound emitted by an airplane travels outward from the point of emission in a circular path. Since the airplane is moving, the center for each outgoing sound wave is different from that of every other wave. These outgoing waves are represented by the circles shown in the figure.

If the aircraft is moving at less than the speed of sound, we will have a situation like that shown at the left. The sound wave emitted when the plane is at A will have passed the point B before the plane arrives. Again, the wave emitted at B will never catch up with the wave emitted at A, so the two waves will never be able to reinforce each other. Someone standing on the ground near the flight path will hear a succession of sound waves arriving at his observation post, and his ear will translate these waves into the kind of steady drone familiar to anyone who has spent time near an airport.

But if the plane is flying at a speed faster than that of sound,

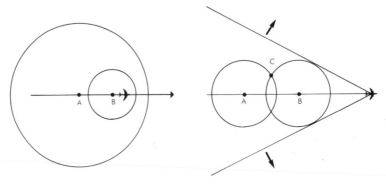

FIGURE 11.6

the situation is quite different. As shown on the right in figure 11–6, the aircraft overtakes the sound wave emitted at A, so that when it emits another wave at B, it is well beyond the circular disturbance due to the first wave. This means that the wave emitted at A and the wave emitted at B can come together and reinforce each other at the point labeled C. There will therefore be a cone of waves trailing backward from a supersonic aircraft, a cone created by the coming together of sound waves emitted by the plane at different points in its flight. Along this cone, the sound waves reinforce each other and build up, just as small stationary waves build up to create the hydraulic jump. Anyone standing in the path of these waves will hear a sudden, sharp sound—the sound known as a sonic boom. These booms can rattle dishes, break windows, and, in extreme cases, crack the foundations of houses.

In a way, a discussion of the sonic boom is singularly appropriate in a book inspired by walking in the mountains. Because of the potential property damage associated with the sonic boom, supersonic flight over the continental United States is restricted to regions of very low population density. Paradoxically, therefore, someone who travels to a distant wilderness area to get away from civilization is likely to hear an occasional boom from military aircraft. From the peaks of the Beartooth Mountains in Montana, I have often watched B-52 bombers and interceptors playing supersonic tag in the almost empty skies.

We can close this discussion of wave reinforcement by mentioning one other familiar situation in which it can be seen. When a boat moves through deep water, it leaves a V-shaped wake in its path, as shown in figure 11–7. This wake is known as a bow wave, and it is present whenever the speed of the boat exceeds the speeds at which waves can travel in the water. The only difference between a bow wave and the sonic boom is that in deep water the speed of a wave depends on its wavelength, while for sound waves in air (or waves in shallow water) all waves have the same speed. The effect of this complication is that a bow wave is somewhat smeared out, and doesn't have the sharp, abrupt nature of the sonic boom. The explanation of the bow wave, however, is exactly analogous to that of the sonic boom. You can test your understanding of all of the phenomena we've discussed by going through the reasoning associated with figure 11–6, but substituting a boat in water for the plane in the air.

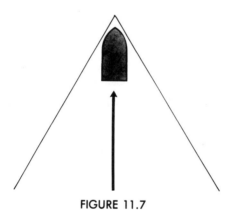

FIGURE 11.7

TWELVE

# Sky Divers and

# Reverse Meteorites

*I chatter over stony ways*
*In little sharps and trebles*
*I bubble over eddying bays*
*I babble on the pebbles....*
*And out again I curve and flow*
*To join the brimming river*
*For men may come and men may go*
*But I go on forever*

—ALFRED, LORD TENNYSON
*"The Brook"*

WE CAN LEARN many lessons from watching the water in a mountain stream, not all of them as subtle and difficult as the hydraulic jump. What makes a set of rapids different from the orderly flow of a river is the presence of obstacles in the stream bed. To get downstream, the water must find its way around a large number of rocks. It is this process that helps to wear down the mountain. At the same time, describing the flow of a fluid around solid bodies is difficult; the problem has been (and remains) a major source of conflict in science.

That particular problem has applications far beyond the description of a mountain stream. In the design of ships, cars, and aircraft it is extremely important to know what effect a fluid, be it air or water, will have on the motion of the structure. "Streamlining" is a commonplace term that refers to the attempt to minimize the force exerted on a moving plane, ship, or car

*182*

by the medium through which it moves; the purpose of stream-lining is to ensure that as little fuel is used to overcome air or water resistance as possible.

The resistance offered by air to objects moving through it is one example of a property shared by all fluids,* a property known as viscosity. Viscosity refers to the effects of friction that occur in fluid flow. From our point of view, there are two important aspects of viscosity: the friction that exists when parts of a moving fluid move against each other, and the friction that exists when a moving fluid runs around a solid. It is the latter that we call resistance, or drag.

We are so used to the existence of friction in ordinary motion that we take it for granted. When you take your foot off the gas and let your car coast to a stop sign, you are letting the friction of the tires against the road slow you down. Because of friction, no motion goes on forever, including the motion of fluids. When you let water into your bathtub, the water eventually becomes still after you turn off the tap, no matter how rapidly it was flowing at first. Like your car coasting to a stop on the highway, the water is eventually brought to a stop by friction.

At first it may be hard to visualize the action of friction within the body of a moving fluid, but you know it must be there. Just think about pouring syrup on a cold morning. Something is slowing down the flow, and that something can only be some sort of frictional effect. The easiest way to picture that effect in a fluid is shown in figure 12–1. Imagine the fluid broken up into a series of thin sheets. When the fluid moves, it is possible for these sheets to slide across one another. A rough analogue to this sort of flow occurs when you take a stack of papers, push it along a table, and then stop the movement with your hand. The upper sheets tend to keep moving. They move forward, rubbing against the sheets below, which are in turn dragged forward and rub against the next lower sheets. The result is similar to what is shown: the overall motion of the stacked plates takes place through the rubbing of each plate against its neighbors. This rubbing is friction, and such is the

*You may be wondering about my use of the term "fluid" to refer both to water and air. In general, "fluid" is used to refer to any system that exhibits flow; it can be used as a generic term applicable both to liquids like water and gasses like air.

effect in the flow of a fluid, which we observe and term the fluid's viscosity. A high-viscosity motor oil, for example, is one in which the friction is high and the fluid motion slow. Cold maple syrup is another high-viscosity fluid. A low-viscosity fluid, on the other hand, flows easily. When a motion of a fluid can be pictured as in figure 12–1, it is said to be laminar (from *lamina*, a thin sheet). The flow of a deep river is a good example of laminar flow.

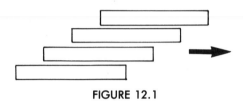

FIGURE 12.1

This model of the flow of a viscous fluid is simple and has been known to scientists for a long time. The historical debate that I referred to earlier arose when people tried to take this simple picture and apply it to the problem of flow around obstacles. It is fairly easy to see that the amount of internal friction in a flowing fluid will depend on how quickly the velocity changes as we move from one sheet to the next. If one sheet is moving over another at a high speed, there will be more friction than if it is moving at a low speed, just as you can warm your hands better by rubbing them together quickly than you can by rubbing slowly. In almost all situations involving water, the overall change in velocity between sheets tends to be fairly small, so we would expect the effects of friction within the fluid and between the fluid and a solid to be small.

The science known as hydrodynamics grew up during the eighteenth and nineteenth centuries using this simple reasoning. The idea that it was legitimate to ignore viscosity in fluid flow was put into more quantitative form as follows: Consider a block of fluid in a moving sheet as shown in figure 12–2. There will obviously be a force arising from friction that tends to slow the block down, a force we have indicated by an arrow in the direction opposite to the direction of the flow. The question of whether this force has to be taken into account depends on how much effect it has on the motion of the block of fluid. Because the block is moving, it has some inertia, some tendency to keep moving and overcome any force tending to slow it down. If

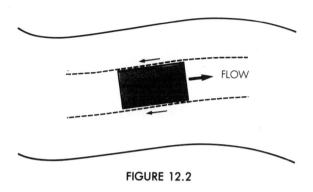

**FIGURE 12.2**

the inertia of the block is very high compared to the frictional force, then the block can, for all intents and purposes, be treated as if it were not subject to friction at all. It would be like a toboggan running over a smooth, icy surface—there would be only a negligible slowing down caused by friction. If, on the other hand, the frictional force is comparable to or greater than the inertia of the block, friction has to be taken into account in all calculations.

Physicists use a shorthand way of talking about the comparison between inertia and friction. They define something called a Reynolds Number (denoted Re), and named after Osbert Reynolds, a nineteenth-century British fluid scientist, as follows:

$$Re = \frac{inertia}{friction\ force}$$

The considerations in the preceding pages come down to this: If Re is much greater than 1, we can ignore friction. Otherwise, we can't.

Now it turns out that for almost every flow involving water, this reasoning tells us that we can ignore friction. This is fortunate, because having to take friction into account almost always makes the problem too difficult to solve without the use of large computers. During the eighteenth and nineteenth centuries hydrodynamics flourished as a theoretical science. People worked out the theory of the tides on the ocean, the theory of waves from beach surf to tsunamis (tidal waves), and the theory of the convection cell, among other fluid phenomena. By and large, these scientists enjoyed great success in their ventures. In a few especially simple cases, such as flow through a cylindrical pipe or around a smooth sphere, they even succeeded in carrying out calculations including friction.

There was only one problem with this massive advance on the theoretical front. Whenever the hydrodynamicists tried to apply their work to what happened when obstacles were placed in actual flowing water, their predictions were wide of the mark. Engineers charged with designing canals, locks, ships, and other structures where forces exerted by water are important finally gave up on the theory and founded a completely separate discipline called hydraulics. This was based almost entirely on the accumulation of experimental results and had a decidedly empirical cast. If you wanted to build a dam, you talked to the hydraulic engineer, but if you wanted to understand the flow of free fluids, you talked to a hydrodynamicist. Unfortunately, if you wanted to understand the flow of water around obstacles in a stream, there was no one you could talk to.

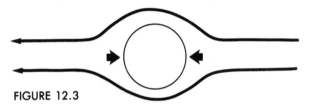

FIGURE 12.3

This state of affairs is best exemplified by a simple problem. Suppose, as shown in figure 12–3, a stream is flowing around a submerged rock—a rock whose shape we will take to be circular. On the average, the relative motions between adjacent sheets of water in the flow are quite small, so the average frictional forces are small. We can therefore neglect the viscosity and calculate the flow of the fluid straightaway. A fluid will exert a force on a solid with or without viscosity—a fact you verify every time you direct a stream from a hose on a sidewalk. The ordinary inertial forces generated by the movement of water in the stream can push on the front faces of obstacles like the rock. The problem is that if the water flows around the rock, as shown in the figure, it also pushes on the back face of the rock, and this tends to cancel out the forces exerted on the front face. In fact, if you work things out carefully, you find that in the absence of viscosity, running water can't exert any force on the rock at all! This result is known as d'Alembert's paradox.

I was particularly pleased to find that this old paradox played such an important part in the history of fluid science because Jean Le Rond d'Alembert has always been one of my favorite historical characters. Born around 1717, the illegitimate son of a cavalry officer and a society woman who maintained a Parisian salon, he became one of the leading physicists and mathematicians of his era, and there are all sorts of theorems and effects named after him. One of my old math professors told us that d'Alembert had been left as a foundling and raised by poor but honest parents. After he rose to a position of prominence and fame, the story went, his biological parents came forward, only to be rejected by d'Alembert, who made it clear that he regarded the people who had raised him as his true family.

I know this story is too good to be true; life is never as simple as all that. Nonetheless, I like it so much that I have steadfastly refused to go to the library to check the facts on d'Alembert's life in the two decades since I first heard it. I'm certainly not going to do so now, so I pass it along as a bit of interesting hearsay.

Whatever the conditions of d'Alembert's life, his conclusions about the force exerted by a fluid on a submerged rock illustrate as well as anything the frustrations of engineers when confronted with the elegant work of the hydrodynamicists. If the water can't exert a force on the rock, how can the rock get carried downstream? Obviously, something is missing here.

The missing ingredient eluded some of the best minds in the human race until the early twentieth century. In 1904, the German scientist Ludwig Prandtl, working at the University of Gottingen, finally resolved the issue and, in the process, healed the rift between theory and experiment in fluid mechanics. We can understand his contributions if we go back to the picture of fluid flow in terms of thin sheets. Suppose a stack of these sheets encounters an obstacle, as shown in figure 12–4 below.

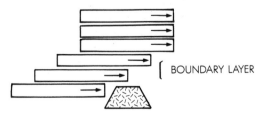

**FIGURE 12.4**

What will happen? If there is friction between the water and the obstruction, as there is bound to be in the real world, the sheet in contact with the rock will come to a halt. The second sheet will then slide over the first, the third over the second, and so on. If, as is the case with water, the frictional effects between the sheets are small, only the first few sheets will slide over each other. Beyond this, the movement of the sheets will be largely unaffected by the obstacle.

You can get a good mental image of this picture of the encounter between a fluid and an obstacle by imagining a stack of well-oiled sheets of metal being moved along. If the bottom sheet hits an obstacle, it will stop, and the first few sheets will slide over each other. The main bulk of metal, however, will continue moving, and the lubricating action of the lower plates will absorb all the effects of the obstacle.

What Prandtl realized was that this picture showed why there had been such problems in understanding flow around obstacles. If we look at the moving sheets of fluid *as a whole*, it is certainly true that the overall motion is one in which there is little relative motion between the plates, and hence little friction. On the other hand, for a few sheets near the obstacle itself, the sheets will be sliding over each other at high velocities, generating a large frictional force. In effect, the region far from the obstacle is what the theoreticians had been talking about—a region where friction could be neglected. Since most of this fluid is in these regions, the overall flow patterns are determined by equations in which friction plays no part. At the same time, near the obstacle the frictional forces must dominate because the relative viscosity between the plates is high. This is the region where the hydraulics people had been concentrating their attention. The great schism, then, came to nothing more than the fact that the two rival groups had been looking at different parts of the flow pattern. It was a classical case of the blind men describing the elephant: each examined a different part, hence each gave a completely different description of the beast. Another way of saying this is to notice that while the relative motion of the sheets may be small in more than 99 percent of the fluid, in the 1 percent of the fluid near the obstacle, the relative velocity of the sheets is very high. But it is this 1 percent that determines the interaction of the fluid with the obstacle.

Prandtl called the thin layer in which frictional forces dominate the "boundary layer." His was one of those ideas in science that are so obviously right and so simple that you think "Of course, that's the way things must be" the minute you hear it. Prandtl must have had one of those "light bulb" experiences when the idea first occurred to him. Once the idea of the existence of a boundary layer caught on, it didn't take long to come to some understanding of the interaction between moving fluids and obstacles. It turns out that this flow around barriers takes on different forms depending on the size of the obstruction and the speed of flow of the fluid. You can probably see all the crucial stages of flow in any stream with a reasonable amount of whitewater in it.

## Laminar Flow

If the flow of the fluid is slow enough and the obstruction small enough, you can get a situation like that shown in figure 12–5. The fluid flows smoothly around the rock, joining up downstream without any hint of ripple or turbulence. The force on the rock due to the friction between it and the water is shown by the arrows. Obviously, including viscosity allows us to get around the d'Alembert paradox, since it allows us to bring the frictional force into the problem, giving a net downstream force on the rock.

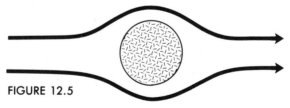

FIGURE 12.5

The flow around a cylinder is one of those simple situations that the hydrodynamicists were able to solve exactly. The same is true of a related problem—the flow around a sphere. In both cases it turns out that the force exerted on the solid by the fluid depends on the fluid viscosity and the cross-sectional area of the solid. As you might expect, the force increases when either of these parameters gets bigger. The existence of the frictional

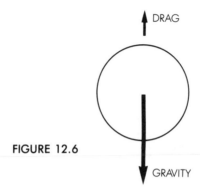

FIGURE 12.6

force on a sphere leads to some interesting consequences for the behavior of things falling through the air. If we concentrate our attention on a small drop, we recognize two forces on it, as in figure 12–6. There is the ubiquitous force of gravity causing the drop to accelerate downward and the force of friction (which we call drag) opposing the downward motion. When a drop begins its fall, it is in the process of being accelerated and its viscosity is small. This means that the drag force associated with the motion through the air is small as well—too small to counteract gravity. In this phase of the motion, the drop will be accelerated downward, but not as rapidly as it would be in the absence of drag. As the drop speeds up, however, the frictional force increases. After a while, the speed reaches the value at which the drag exactly balances the force of gravity. At this point, the acceleration stops and the drop falls at a constant speed, a speed known as the terminal velocity. An object falling through the air, in other words, eventually reaches a limiting velocity, and it traverses the rest of its path at that particular velocity. This is true for small raindrops, where the air flow is smooth, and it is also true for larger bodies (such as sky divers) where turbulence becomes important. We'll discuss this case later.

Bird hunters, who use shotguns loaded with buckshot, often have the experience of being "rained on" by their own ammunition. What happens is that one member of the party fires into the sky, and the shot, after moving upward until its momentum is spent, falls back down. From our discussion of the raindrop, we know that no matter what speed the shot has when it leaves the barrel of the gun, once it starts to fall, it will

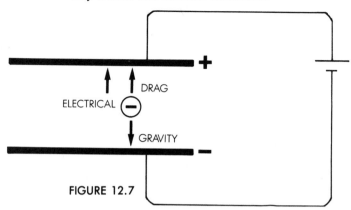

ELECTRICAL

DRAG

GRAVITY

**FIGURE 12.7**

eventually reach the terminal velocity—a velocity too small to cause harm if it hits something. Had the gun been fired in an airless environment, on the other hand, the shot would return to the ground at the same velocity as that with which it left the gun barrel. There are a lot of hunters who owe their continued good health to viscosity!

A more scientific application of the concept of terminal velocity was made in 1910 by the American scientist Robert Millikan when he measured the charge on the electron for the first time. A sketch of Millikan's apparatus is shown in figure 12–7. The idea is quite simple. A small drop of oil is exposed to radiation so that it acquires an electrical charge. The drop is then allowed to fall through the air between large plates connected to the poles of a battery. The forces on this drop are gravity and drag, as was the case with the raindrop, and the upward electrical force associated with the plates (the attraction between the negatively charged drop and the positive pole of the battery). The progress of the drop is monitored with a small telescope and its response to the forces is measured. The electrical force is varied by changing the voltage on the battery, the motion is measured; the charge on the drop is then changed with another blast of radiation and the motion is again monitored; and so on. Eventually, enough data are accumulated to show that the charge on the drop is always an integral multiple of some basic unit. It is this basic unit of charge that Millikan identified as the charge on the electron.

More recently, the Millikan technique was updated with modern technology and used to search for quarks, the particles we believe are the basic building blocks of the atomic nucleus.* In most modern experiments, the drop falls in a frictionless vacuum, so drag doesn't enter the picture.

## Separation

One way of picturing the flow around the cylinder in figure 12–5 is to imagine that the water encountering the front is accelerated on the upstream rock face until it reaches the apex, then decelerated on the back face until it is ready to rejoin the main flow. So long as the frictional forces are small, this is exactly how the flow behaves. When you look at the rock in the stream, you see no downstream ripples or bubbles; just a smoothly flowing current.

As the flow velocity increases, the inertia does also. If the flow is fast enough so that the Reynolds number is greater than 1 (i.e., if we get to a situation where inertia dominates friction), the character of the flow changes dramatically. As the fluid is accelerated on the upstream face of the rock, some of its energy is dissipated as heat because of friction. Consequently, when the fluid encounters the decelerating forces on the downstream side, it doesn't have enough energy to overcome them. The situation is analogous to what you get if you roll a ball into a hollow. If there is no friction, the ball will roll down to the bottom and then up the far side to a height equal to the one from which it started. But if there is a lot of loose sand lying around, so that frictional forces are high, the ball will not be able to roll very high on the opposite side. Just as the ball rolling downhill is accelerated by gravity, the fluid on the upstream face of the rock is accelerated by pressure. Just as the presence of friction prevents the ball from regaining its original height, it prevents the fluid from regaining its original velocity on the downstream side of the rock.

When this happens, the fluid in back of the rock can actually reverse direction, as shown in figure 12–8. Small eddies separate from the main flow—an effect you can easily observe in a stream. For future reference, we should note that this sort

*A detailed description of the search for quarks is given in my book *From Atoms to Quarks* (Scribners, 1980).

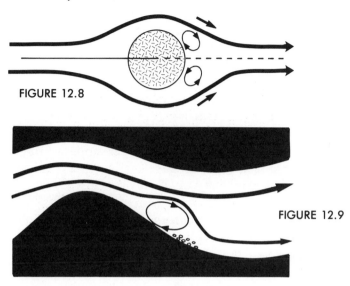

FIGURE 12.8

FIGURE 12.9

of flow pattern usually occurs when the Reynolds number is above 4.

You get the same effect in the situation shown in figure 12–9 where a stream is suddenly constricted. This type of eddy is probably much more familiar to you. Since the fluid tends to stagnate in regions where the flow is reversed, debris being carried by the stream often accumulates in the eddies, making them visible even to the casual observer.

There are many places other than streams where the formation of eddies is important. Perhaps one of the most unexpected is in medicine, where researchers worry about the causes of hardening of the arteries, arteriosclerosis. One model for the disease works this way: Blood flows down an artery (which we assume can be represented by a cylindrical pipe) and encounters a small constriction, just as the stream in figure 12–9 does. On the downstream side of the constriction, eddies form and debris collects. The fact that the flow in the stagnation region does not cleanse the arterial wall means that complex chemical processes can take place in a relatively unperturbed environment. Consequently, a hardened layer of fatty deposits builds up, further restricting the flow of blood. The further restriction leads to the formation of new and larger eddies, which in turn lead to more deposits. It's not hard to see that this process, building on itself, will eventually clog the artery completely.

In point of fact, the only differences a theorist sees between eddies in arteries and eddies in open streams are relatively minor ones: the difference in properties between blood and water, and the difference between the rigid banks of a stream and the elastic walls of the arteries. Otherwise, the two problems are the same, so that understanding what happens when a stream encounters obstacles may eventually lead to better modes of treatment (or prevention) for arterial disease.

## Shedding Vortices

As the speed of flow increases further, the Reynolds number increases. When the Reynolds number gets up to about 40, something new happens. The eddies behind the rock begin to detach themselves and be swept downstream. A line of vortices forms behind the rock, as shown in figure 12–10. We say that the vortices are "shed," first from the right-hand side, then from the left. The result is known as a von Kármán vortex trail, after Theodor von Kármán, the Hungarian-born pioneer in fluid mechanics whose work on turbulent flows at Cal. Tech. in the 1920s and 30s laid the foundations for the modern science of aerodynamics.

You have undoubtedly encountered von Kármán vortices in your life, although you may not have realized what they were. Next time you're driving along the Interstate and a big semi truck passes (as they usually do), pay attention to your steering wheel. You'll notice that as the truck pulls up alongside of you, your car seems to be pushed toward the side, away from the

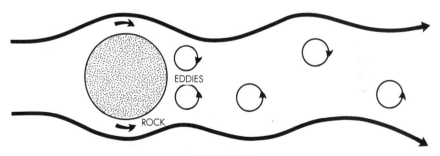

FIGURE 12.10

truck. After the truck passes, your car is pushed first one way, then another. There may be four or five such shoves before the truck gets far enough ahead for its effects on the air to vanish.

The forces you feel are shown in figure 12–11. You feel the first sideways push when you enter the slipstream created by the air being pushed aside by the truck (point A). The buffeting after the truck passes, on the other hand, is caused by your car encountering the vortex trail. At point B you are pushed to the right, at point C to the left, and so on.

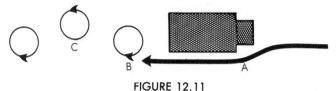

FIGURE 12.11

When the vortices are shed, a force is exerted on the body doing the shedding. If the shedding is timed just right, this can have serious implications for large structures in the wind. It is well known that bridges (to take one example) are susceptible to damage if subjected to impulses at regularly timed intervals— this is why soldiers are always told to break step before marching across one. In 1940, a newly built bridge over the Tacoma Narrows in Washington State was destroyed when gusting winds started it swaying. I always enjoy showing films of the disaster when I teach a class of engineers, just to keep them from getting too cocky.

On the more prosaic level, when you hear a telephone or electrical wire "singing" in the wind, you can be sure that the wire is being "plucked" by the forces exerted by the shedding of vortices as the air flows by. You can also "see" von Kármán vortices in the swirling leaves in the wake of a car traveling down a street on a dry autumn day.

## Turbulence

As the velocity of the fluid increases, the basic vortex structure we've described persists, but it becomes increasingly mixed with a general turbulence. When the wake downstream of a rock in a swiftly moving stream appears to be nothing but a froth of bubbles and foam, you may well wonder where the vortices are. If you were able to watch a slow-motion film of the wake, you would definitely see pieces of debris moving in the familiar vortex pattern, but the overall motion would be combined with a kind of random jittering.

In a stream you are unlikely to see Reynolds numbers above a few thousand, so the catalogue of fluid phenomena we've compiled so far is pretty much what you will be able to observe on a hike in the mountains. If, however, you were charged with designing an airplane or a space vehicle, you might have to consider situations where the flow of air around your craft was characterized by Reynolds numbers in excess of a million. At such high speeds, even the boundary layer becomes turbulent, and the wake looks like the one in figure 12–12. The speed is so high that the boundary layer is blown off the front part of the sphere, and a turbulent wake (with some underlying vortex structure) trails along behind. If the speed of the sphere is high enough, the frictional effects can heat it up considerably, a fact which explains why it is so important to have heat shields on the space shuttle during reentry.

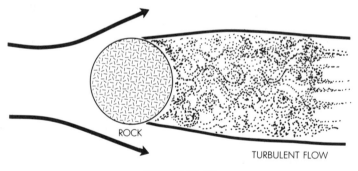

ROCK

TURBULENT FLOW

**FIGURE 12.12**

## Sky Divers and Reverse Meteorites

Once we have an understanding of how fluids interact with obstacles in their path, or, what is equivalent, how solid objects move through a fluid, we can look at some interesting applications. We saw, for example, how a raindrop falling through the air reached a terminal velocity where the downward pull of gravity was balanced by the frictional forces. In that case, we had laminar flow around the falling body—no wake, no turbulence.

For larger objects or for higher velocities, we begin shedding vortices, and the mechanism by which energy is transferred from the falling body to the air changes. For laminar flow, this mechanism is just ordinary friction—the rubbing of the air against the solid. As soon as we get vortices, however, another energy transfer enters the picture: the motion of the solid can be converted either into heat (as with laminar flow) or into the swirling motion of the fluid in its wake. Once we get past laminar flow, energy losses due to drag increase quickly. In laminar flow, doubling the velocity of the object results in a doubling of the energy loss, but when the vortices start to form, things get worse. Doubling the velocity for high Reynolds numbers results in a quadrupled energy loss.

This means that the terminal velocity for large and small objects can be quite different. For example, if a falling raindrop is as small as a fine grain of sawdust, the flow of air around it can remain laminar. In this case, its terminal velocity is about 3 miles per hour—the speed of a brisk walk. But if the raindrop grows even to the size of a grain of salt, it will start shedding vortices. By the time the raindrop has grown to a size you would notice, it would be shedding a complex turbulent wake as it fell. Its terminal velocity would then be about 20 miles per hour. A sky diver, on the other hand, is always so large that he or she will shed vortices right from the start. Consequently, there is no question about the terminal velocity involved in the fall. It will be somewhat higher than 100 miles per hour, depending on the diver and the position assumed during the fall.

Meteors entering the atmosphere present a somewhat different case from falling rain or parachutists. Meteors start out with a very high velocity and are slowed down by friction as they fall, but they almost never slow down to the theoretical terminal velocity. They usually either burn up or hit the earth before they slow down appreciably.

Actually, what interests me about high-speed bodies in the atmosphere is not so much the meteor falling from space, but the reverse process—projectiles launched from the earth into space. I call these reverse meteorites.

Every payload that is launched into space today is carried by a rocket. This mode of launch suffers from a profound disadvantage. If the rocket is going to burn a certain amount of fuel when it is a mile above the surface, it has to carry that fuel with it on its ascent. This results in enormous inefficiencies. For example, if we wanted to lift a 1-kilogram mass to an orbit normally taken by the space shuttle, the amount of energy we would have to expend in principle is only about eight kilowatt-hours—enough to run an air conditioner for a day. If you bought this amount of energy from your local utility, it would cost less than a dollar. Compare this with the enormous cost of a shuttle launch (over $1000 for the same mass) and you have a good idea of what I mean when I talk about the basic inefficiencies of the rocket.

The Princeton physicist Gerard O'Neill has proposed an alternative means of moving cargo into space, a device called the mass driver.* In essence, a mass driver is a stationary device that accelerates a payload to a high velocity, using a power source which stays on the ground and is itself not accelerated. Once brought to a high speed, the payload moves out into space of its own accord—a meteorite in reverse. In this scheme, the energy need be applied only to the payload; the fuel remains on the ground.

As initially conceived, the mass driver was to be used on the moon. Since the moon has no atmosphere, there is no air resistance to slow the payload down once it is launched. Originally, it was believed that the presence of an atmosphere would prevent the use of mass drivers on the earth. But if you think about it for a moment, you'll realize that this reasoning is flawed. Meteorites, reentry vehicles, and space shuttles routinely survive the trip from space to the earth's surface. Why couldn't we design a vehicle that could make the trip in reverse? The only problem that an earth-based mass driver might encounter would be atmospheric friction and drag. If this effect were large enough, it might simply be uneconomical to supply the energy needed to overcome it, in which case we'd be back where we started.

A few years ago, Ray Cheng (then an undergraduate at the University of Virginia) and I did some simple calculations on

*This device, along with many other aspects of future space technology, is discussed in O'Neill's excellent book *The High Frontier* (Morrow, 1976).

this problem. Our result: even in the worst-case scenario, assuming no attempt is made to streamline the payload vehicle, the extra energy needed to overcome frictional effects in the atmosphere amounts to a factor of 3 over the energy required for the case where there is no atmosphere. In other words, if we take the atmosphere into account, the energy cost of lifting a kilogram to shuttle orbit goes from eight to twenty-four kilowatt-hours. At 6¢ per kilowatt-hour (the present cost of electricity in Virginia), the cost goes from about 50¢ to about $1.50. Either way, it's an enormous savings over rockets. More realistic calculations, including the effects of streamlining, indicate that the added energy needed to overcome friction is only around 10 to 20 percent, well below our worst case result.

It's no wonder that space enthusiasts talk about the time when it will be cheaper to send a letter into space from Cape Canaveral than to mail it to New York.

Not to mention quicker.

# THIRTEEN

# The Road to Chaos

*chaos (kā'os') ... Any condition or place
of total disorder or confusion....*

—*American Heritage Dictionary*

THE PROGRESSION OF a mountain stream from a barely discernible ripple to whitewater is an example of what is called the "transition to turbulence." The wild, chaotic movement of swiftly flowing water is one example of a type of process whose study constitutes one of the major areas of research in the physical sciences. For although it may seem to be a contradiction in terms, the study of "chaos" and chaotic systems today occupies a large number of scientists, and the results of their work are already being applied in disciplines as far apart as aircraft design and field ecology. The sorts of things we've already seen happening in the mountain stream offer a vivid example of the study of chaos.

Most simple physical systems exhibit what is called linear behavior. If you hang a weight on a spring, the spring will stretch. Double the weight and you double the stretch. In the same way, when you turn the volume knob on your stereo, you get a certain level of sound. If you turn the knob twice as far, the volume doubles. In smooth laminar flow around an obstruction, we get the same sort of thing: the force exerted on the obstruction doubles if the velocity of the flow doubles.

In all of these instances, there is a point beyond which the simple linear behavior disappears. For the spring, this point is reached when the weight causes the spring to stretch beyond its elastic limit. From that point on, the spring never comes back to its original condition after you remove the weight; small increases in weight will cause the metal to stretch very far indeed and even to break in half. In the same way, your stereo will reproduce the sound input faithfully unless you turn the knob up too high. Then the sound starts to become distorted and is no longer an accurate rendition of the input: the response of the amplifier has stopped being linear. For the water flowing around an obstacle, the end of the linear regime comes when eddies start to form. From that point on, as we saw, doubling the velocity causes the force exerted on the obstacle to grow much more quickly than it did for simple laminar flow.

In all these examples, we are dealing with a system that first responds one way to changes in conditions, then suddenly starts to respond in quite a different way. In the language of physics, such systems are said to be nonlinear. From the time of Newton until quite recently, there has been a sort of gentleman's agreement among scientists that nonlinear systems should be politely ignored. The reason was simple: the mathematical equations that describe nonlinear systems are notoriously difficult to solve. Except in a few instances, where some fortuitous circumstances allowed someone to find a mathematical trick that made it possible to find easy solutions for a nonlinear system, scientists confined their attention to linear systems—systems that they could solve.

This is standard operating procedure in every area of science that I know about. The general philosophy is that you describe the system as best you can within the limits of what can be solved mathematically. The stretched spring we talked about, for example, can be described by very simple equations in the linear regime, so that's where springs have been studied. Once the spring reaches its limits and stretches irreversibly, the problem of describing it slips from the realm of the pure scientist to that of the engineer. The split between hydraulics and hydro-dynamics which we explored in chapter 12 is not an isolated example of this sort of divorce between the theoretical and the empirical in science.

I remember being surprised during my graduate student days by the number of times visiting speakers would comment on the fact that their models of elementary particle processes were poor or incomplete descriptions of nature, but that they could be solved while more realistic models could not. The standard joke used to make this point was about a man looking around under a streetlamp. Someone came up and asked what the trouble was, and he replied that he had lost his key. Both men search for a while, then the second asked: "Are you sure you lost it here?"

"No, I lost it back in the alley."

"Then why look under this lamp?"

"There's more light here."

As in so many areas of life, you do the best you can with what you have.

The development of large, fast computers during the past few decades has led to a dramatic change in the situation as far as nonlinear systems are concerned. Ability to mask the com-plexities of previously insoluble equations under sheer com-puting power has produced a generation of scientists who have been able to extend the limits of the calculable far into the region that used to be considered terra incognita. In the process, we have discovered some surprising things about the way that com-plicated systems work.

Let's take turbulence in moving water as a case in point. If you watch the whitewater downstream from an obstacle in a swiftly moving stream, you'll see something like what is shown in figure 13 – 1. Immediately behind the rock are large eddies, as we discussed in chapter 12. Farther downstream, these eddies break down into smaller eddies, and these smaller eddies break down into still smaller ones. This process eventually gets the size of the eddies down to the point where collisions between the atoms in the swirling vortices and the stream itself cause the former to lose their excess energy. At this point the eddies disappear to merge again with the bulk of the moving water. The process by which energy is transferred from large eddies to progressively smaller ones until the initial energy of the large ones is absorbed into the general fluid motion is seen in all sorts of fluid systems, from mountain streams to the motion of gases in the stars.

Suppose we put a small bit of wood on the water at the point labeled A in figure 13–1. It will follow some complicated path through the eddies and eventually emerge at the point labeled A'. Suppose we put another bit of wood in the water at the point labeled B, near point A. It, too, will work its way through the rapids, although its path will probably not be anything like that followed by A. In general, the bit of wood that starts at B will emerge on the far side at some point B'. The fact is that no matter how close the starting points A and B are to each other, by the time the bits of wood have worked their way through the turbulent water, they will have separated, and the points A' and B' will, in general, be far apart.

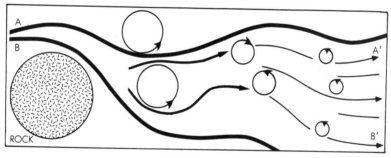

FIGURE 13.1

A system in which the final results can be very different, even though the initial conditions are nearly the same, is said to be "chaotic." The connection between nonlinear systems and chaotic systems is far from simple. All chaotic systems are nonlinear, but there are many nonlinear systems that are not chaotic, and we shall look at some of these later on. For the moment we simply note that a system operating with the implication that two points close together at the start will not be close together at the finish is a system taken to represent "chaos."

The word seems justified by common experience. For example, if you drop a ball from a rooftop, it will hit the ground somewhere near the building from which you throw it. If you move over a few feet and drop the ball again, it will hit the ground close to where it landed the first time. In this homely example, changing the initial conditions (the point from which the ball is released) by a small amount changes the outcome (the point where the ball lands) by a small amount as well. The dropped ball is an example of a nonchaotic system, and it is a lot easier to deal with this sort of problem than with the least complicated example of chaos.

But ease of solution aside, chaotic systems can be used to clarify our thinking about an old question in natural philosophy—the question of whether events in the world are absolutely determined. In its simplest form, the doctrine of determinism holds that if we know the present state of the universe, then by the judicious application of the laws of nature we should be able to predict its state at any future time. Pierre Simon de Laplace, a physicist and mathematician of the late eighteenth century, put the idea best when he talked about a "divine calculator." He argued that if you knew the position and velocity of every piece of matter in the universe at some instant of time, and if you were good enough at solving the equations that describe motion, you could predict the position and velocity of every such piece at any time in the future. Laplace's model for this was the Newtonian solar system, in which planets interacted with each other and the sun through the force of gravity. The equations that describe the gravitational force are simple, and Laplace felt that they would probably be solved completely at some point. (This expectation wasn't fulfilled until the advent of the computer. Even the solar system is too complicated a system to yield easy answers to the unaided mind.) Regardless of the question of the feasibility of the divine calculator, the idea that all future states of the universe are predictable in principle exerted a powerful force on the philosophers of the Age of Reason. To take just one example, think of the implications of this idea for the doctrine of free will. Can a human being really be said to be free if his position and velocity at every future time can be written down on some sort of computer printout?

It is usually argued that the advent of quantum mechanics has, at the very least, rendered Laplace's argument moot. We now know that the laws governing the behavior of atoms are such that it is impossible to determine the position and velocity of even a single atom at any instant of time, much less the positions and velocities of every atom in the universe. Nevertheless, Laplace's divine calculator remained an interesting "what if?" question within the confines of the Newtonian clockwork universe. The idea is that deterministic systems can exist only if we ignore the underlying laws of quantum mechanics, which is what we usually do in describing ordinary events in the world. So long as we have a set of equations, preferably equations we can solve, the world can be deemed determined as Laplace believed it to be.

Unfortunately, even this version of the divine calculator does not survive when we look at chaotic systems. To see why this is so, look again at figure 13–1. The laws that govern the flow of fluids are quite simple—deceptively so. In principle, we could use them to predict the location of the points A' and B' if we were given the locations of A and B. But even if we had a divine calculator to make these predictions for us, it would do us little good. Suppose we started with a string of wood chips lined up solidly between A and B. The nature of chaotic systems is such that knowing what happens to a chip at A tells us nothing about what happens to a chip near A. No matter how close together we bring the points A and B, the points A' and B' can turn out to be arbitrarily far apart.

The full significance of this fact will become clear if you think about the act of measuring the position of the spot we have labeled A. We might, for example, specify the position of the piece of wood chip we place there by giving its distance from the bank of the steam. But no matter how we specify the point, what we are doing in effect is using a yardstick to locate the starting point for our theoretical prediction. And no matter what sort of yardstick we use, there is always some limit to the accuracy with which we can specify this position. Take an ordinary yardstick. With it we might be able to determine the position of A to within one-eighth of an inch. We could then say something like "The point A lies between 18¾″ and 18⅞″ from the bank." If we used more precise instruments, we might be able to reduce this uncertainty to thousandths of an inch or even less; but in principle there will always be a smallest range—a highest level of accuracy—beyond which we cannot specify the starting position. This means, as shown in the figure, that the point A can lie anywhere within a shadowy area determined by the accuracy of our measurement.

Now think back to our discussion of chaotic systems. We know that if two chips of wood start at neighboring points in this shadowy area, they can wind up very far apart at the far end of the turbulence. This is illustrated by the paths starting at C and D in figure 13–2. We have just found that we can never tell—even in principle—whether the chip starts at C or D. More precisely, when we look at the starting point of the chip, we cannot tell whether its distance to the bank corresponds to C or D, hence we are forever uncertain about what number to put into our equations when we start what we hoped was a certain prediction. So, if we give the divine calculator the position corresponding to C, an answer corresponding to a point like C' will come out; while if we give it the position corresponding to D, it will get D'. It is in the very nature of chaotic systems, in other words, that we cannot predict the outcome of a given experiment, *even when we know the equations that govern the system exactly.*

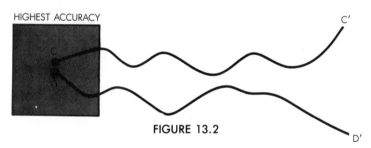

HIGHEST ACCURACY

C'

D'

FIGURE 13.2

This fact was first realized in the 1950s by the meteorologist Edward Lorenz, who was thinking about predicting the weather. So far, we have been using the flow of water to illustrate chaos, but there are all sorts of chaotic systems in nature, and the underlying principle in each is the same as for water. Predicting the weather is a much more complicated task than predicting positions of wood chips in a stream, of course, if only because of all the parameters—temperature, barometric pressure, wind velocity, etc.—that go into a designation of the state of the weather at any given point. Still, we can make the analogy with flowing water by thinking about the way the temperature at a given point evolves with time. In figure 13–3, we show such a graph. The temperature starts out at the value $T_1$ when we

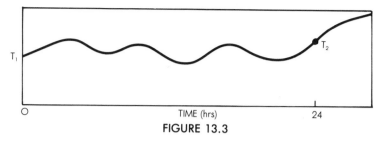

FIGURE 13.3

start the clock (at time = 0). It fluctuates up and down, finally reaching a temperature $T_2$ twenty-four hours later. The similarity between the "flow" of the temperature through time and the flow of the fluid in the stream should be obvious, and the visual similarity in the figures is matched by the mathematical similarity between equations that describe the two curves.

By analogy with whitewater, we can argue that had the temperature at time zero been *slightly* different from $T_1$, the temperature a day later would have been *very* different from $T_2$. The weather, like the turbulent stream, is a chaotic system; the actual prediction of the weather at a given location involves much more than is shown in figure 13-3. Not only do we have to know what the weather is at the location now, we have to know also what the weather was at several other locations. Weather in the United States moves from west to east, so to predict the weather in New York, we may have to know the present weather in Buffalo, Cleveland, Chicago, and so on. This complicates the calculation, but the result is the same. The atmosphere is a chaotic system, just like whitewater.

What Lorenz realized was that the conditions of the atmosphere could not be predicted very far into the future, for exactly the reasons that we found we couldn't predict the motion of a chip of wood in whitewater. To predict weather we would not only need infinitely precise measurements of multiple conditions, but we would need them at every point in space as well. In effect, we would have to cover the earth with infinitely precise weather stations in order to predict the weather, infinite precision not being possible in any case.

Having said this, I should point out that it takes a certain amount of time for the chaotic nature of any system to manifest itself. The two wood chips in figure 13-2 do not separate immediately; they move apart slowly as they flow downstream. In just the same way, the atmosphere does not often change drastically from one hour to the next, and it is this relative slowness of change that allows us to predict the next day's weather with some degree of success. It is only when we start talking about the long term—time scales of weeks and months— that the chaotic nature of the atmosphere becomes evident and the ability to predict vanishes.

The weather service does publish long-term forecasts for three-month periods, but its staff does not arrive at these forecasts by observing the present weather and using computers to solve equations to predict future states of the atmosphere. The forecasts are derived from what is basically the old *Farmer's Almanac* technique. A lot of information about the past few decades' weather is fed into the computers, and the predictions are based on what the weather has done in the past when the situation was similar to the present one. This sort of prediction does not presuppose any detailed understanding of the working of the atmosphere; it is just the ability to correlate large amounts of data gathered from the past. It corresponds to throwing many chips of wood on the stream and seeing where they wind up, then predicting the course of a new chip by likening it to the ones that have gone before. The system works, of course, but not all that well. The last time I checked, the *Farmer's Almanac* had a slightly better track record than the weather service in its long-term forecasts.

Another field in which the theory of chaos is beginning to play a role is in the study of populations of animals and plants. Observers have noticed seemingly chaotic behavior in populations in the field: one species may go through a population explosion, another crash, yet another stay pretty much the same. At a loss to explain such wild variations in the data, ecologists have tended either to follow the lead of an earlier generation and ignore the chaotic data, or to argue that the existence of this sort of behavior means it is impossible to write down simple laws for the evolution of populations. Just a few years ago, I heard a prominent ecologist argue in public that such behavior in populations proved there is no such thing as a "balance of nature." The lesson of the mountain stream shows that neither of these views is valid. Chaotic systems exist in nature, and we shouldn't be surprised to see them in populations. At the same time, the existence of chaotic behavior does not imply that the underlying laws of nature must be complex; indeed, later in this chapter we present a very simple "law of nature" that leads to equations whose solutions are chaotic. The laws that govern the behavior of water flow in a turbulent stream are not difficult to write down; they are quite simple in principle. Yet they lead to behavior far more chaotic than that observed in populations of animals in the field. What this means is that quite possibly laws just as simple will be discovered to explain population biological systems, although these laws will, like those of fluid dynamics, predict chaotic outcomes.

Having examined the idea of chaotic systems, seen how they occur in nature, and noted some of their implications, the only problem left is to try to understand exactly how it is that systems that behave simply under one set of conditions can behave chaotically under others. Perhaps one of the easiest systems to picture is one involving a population of insects that lives during the summer and whose eggs survive over the winter. In such a population, the number of insects that start the summer next year will depend on the number of insects that are alive at the end of this summer.

In the simplest possible case, each pair of insects will produce a certain number of eggs, and some fraction of these will actually hatch in the spring. If this were all there was to the story, the insect population would follow a simple linear relationship between the number hatched next year and the number hatched this year. Double the number of insects present today and the number hatching next year will double as well.

But, as Malthus pointed out, populations cannot simply grow without limit. If the food supplies available to the population are finite, then not every insect that hatches this year will survive to lay eggs to hatch next year. This means that in the real world, doubling the number of eggs hatching now does not result in doubling the eggs hatching next year. In the language we have been using, including the effects of the food supply on the insects makes the relation between this year's population and next year's nonlinear.*

*If we call $n_1$ the number of insects alive now and $n_2$ the number that will hatch next year, a simple nonlinear equation relating the two might be
$$n_2 = E\,n_1 - C\,n_1^2$$
where E is related to the number of eggs laid and C determines how many insects die from the effects of competition.

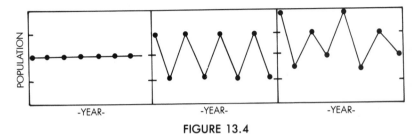

FIGURE 13.4

The simplest situation we can have is shown on the left in figure 13–4. The population reaches some sort of equilibrium and then stays there. The number of insects in each generation is the same, and there is no variation from one year to the next. So long as the insects practice some sort of birth control and don't lay too many eggs, this kind of stability can be maintained.

But birth control, as the human race is learning to its sorrow, is difficult to achieve. Suppose, for the sake of argument, that one year the insects lay an unusually large number of eggs. Then the next year there will be too many insects hatched, and a large number will die off. This will result in fewer eggs the year after. But in that year, the small number of insects hatched would survive more easily, reproduce happily, and leave a large number of eggs. The result would be a population like the one shown in the center of the figure. Large and small populations would alternate on a yearly basis.

If the number of eggs laid increases beyond that necessary to produce this effect, you could get a situation like that shown on the right. A large number of insects would hatch one year, but the fierce competition would cause the population to drop drastically, resulting in very few eggs the following year. This would result in a high population the third year; but, because there were so few insects around in the second year, the third year's population would not attain the heights reached the first year. The result is a population that repeats a cycle every four years instead of every two.

If we raise the number of eggs beyond the number that initiates the four-year cycle, we soon find ourselves with an eight-year cycle, then a sixteen-year cycle, and so forth. Eventually, the cycles get so complex that we have a situation like the one shown in figure 13–5. The population jumps around from year to year in what appears to be a completely random way.

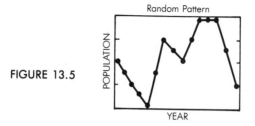

FIGURE 13.5

In this situation, what we observe is a population which in any one year is wildly different from the population in any other year: the situation is chaotic. Yet the system is perfectly determined. I can tell you precisely how many insects there will be in any year provided you tell me *exactly* how many there were the year before, how many died, and how many laid eggs. But if you miss any of these numbers, even by so much as one egg laid by one insect, the predicted population will be wildly different from what you actually see. Such is the nature of chaos.

There is a way of looking at chaos that you may find useful. We saw that the insect population approached chaos by going through a series of fluctuating population states. First, the stable population repeats itself every year. Then we encounter a two-year cycle, a four-year cycle, and so on. In an actual population, we would expect these cycles to be mixed in together. If the number of eggs laid by some insects could be high enough to initiate an eight-year-cycle, for example, we would expect others to lay enough to initiate four-, two-, or one-year cycles. If we plotted the number of insects laying eggs corresponding to each cycle, we might get something like the graph shown in figure 13−6. As the number of eggs laid increased, more and more cycles would be included (sixteen, thirty-two, sixty-four, etc.), until finally we would have a situation in which all possible cycles are present at the same time. This corresponds to the chaotic state of the system, and we say that the insect population has followed *the period doubling road to chaos.*

The stages by which a mountain stream approaches full turbulence also shows some characteristics of the period doubling road to chaos. For laminar flow or the simple eddies shown in figure 13−7, the flow at any downstream point is steady in

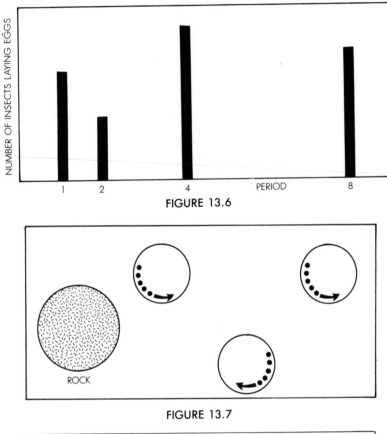

FIGURE 13.6

FIGURE 13.7

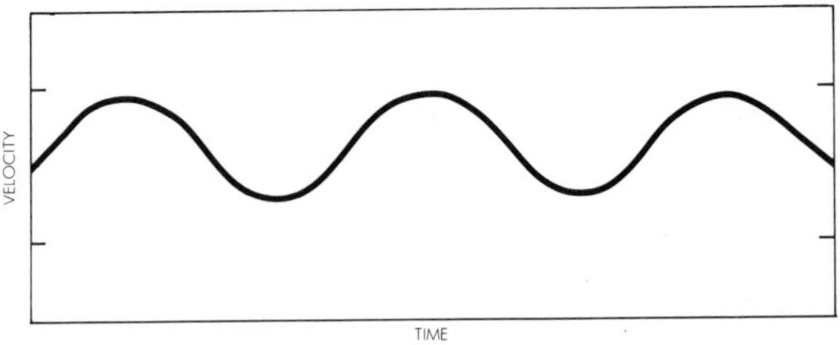

FIGURE 13.8

time. This corresponds to the steady insect population in our example. The appearances of von Kármán vortices as shown in figure 13–8 causes the velocity of the fluid at some downstream point to increase and decrease regularly, as shown on the left. This corresponds to the two-year cycle in the insect

212

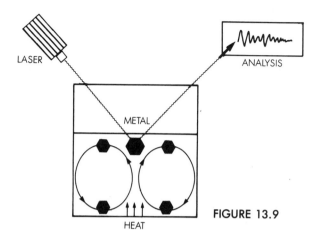

LASER

ANALYSIS

METAL

HEAT

FIGURE 13.9

population. As the downstream vortices start to break down into smaller and smaller eddies, more and more periods are added until we have fully developed turbulence.

A third place where this sort of process can be found is the development of convection cells, such as those seen in boiling water (see chapter 1). Since most of the motion of the water takes place beneath the surface, you need special equipment to see what's going on. A typical experimental apparatus is shown in figure 13 –9. Small filings of metal are dropped into a boiling fluid and a laser is shone in. By analyzing the light reflected from the metal (which moves with the water), we can deduce its motion in much the same way as a policeman can tell the speed of cars on the highway using radar.

The results of this sort of experiment are easy to summarize. As the temperature at the bottom of the fluid is increased, convection cells of the type we discussed on pages 16–18 appear. These correspond to a regular motion of the fluid, a motion with a fixed period. They correspond to the two-year cycle in insect population or the von Kármán vortices in a stream. As the temperature is increased, another motion is superimposed on these convection cells, corresponding to vortices rotating half as fast as the basic convection cells. These vortices, corresponding to the four-year cycle in the insect population, represent the first period doubling in the convection system. As the temperature is increased still, more period doublings occur until, as in the other examples we've discussed, we arrive at full chaos.

Before leaving the subject of chaotic behavior, I want to make a few points. First, although period doubling is a common road to chaos, it is not the only one found in nature. The others are equally predictable, but somewhat more complicated and difficult to describe. Second, the actual transition from a system where many period doublings have occurred to a fully chaotic state, where all periods are present, is much more complex and poorly understood than I have indicated. But this is just what you'd expect in a field that is at the frontier of science.

The fact that systems as different as insect populations, a flowing stream, and a boiling fluid all show the same characteristic behavior as they approach chaos is as good an argument as I can give that nature is governed by a small number of general principles and laws. Apparently there is a kind of order, even in chaos.

# FOURTEEN

# The Mysterious
# Twisted Trees

*With right and left all works are done*
*And the Lifesman knows them, every one*

*—Plainsong of Perkins, Gent.*

I T WAS A HIKE dedicated to exploration. We thought we
would just be following an unfamiliar branch of the trail
to wherever it led; but before we were done, we found
that we had embarked on a voyage of the mind as well.

My companion was my long-time friend and fellow physicist,
Jeff Newmeyer. He started his professional career as a theo-
retician, deep in thought about magnetic monopoles and quantum field theory. Deciding that the otherworldliness of this kind
of science wasn't to his taste, he moved into the aerospace
industry, where basic research, public policy, and practical engineering are uneasy bedfellows. One purpose of our trip into
the mountains was to take Jeff's mind off the problem attendant
upon his role as a major spokesman and liaison for his company
in work on the Strategic Defense Initiative ("Star Wars") program.

As we worked our way up from the trailhead and approached
the timberline, the trees began to thin out a little. It was at that

A tree growing with a right-handed spiral. If you slide the fingers of your right hand along the grooves, your hand is carried toward the top of the tree. Custer National Forest, Montana.

point that we first noticed a rather unusual phenomenon. A large tree had died and, over a period of several years, its outer covering of bark had fallen off, revealing the wood underneath. In this dry northern climate wood rots slowly, so the tree trunk itself was fairly solid. We noticed that the wood seemed to be arranged in a helix (see the photograph above), as if the tree had grown upward like a corkscrew. In fact, we noticed that the wood in the tree appeared to have grown in the direction of a right-hand screw.

*216*

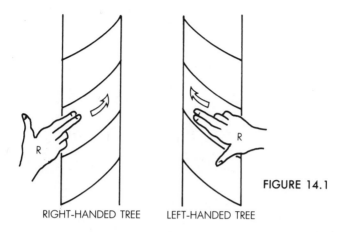

RIGHT-HANDED TREE        LEFT-HANDED TREE

FIGURE 14.1

This "right-hand" business is physicist talk. Since the concept plays a crucial role in what follows, we might as well go into it here. Suppose, as in figure 14–1, that you place your right hand with the fingers along the grooves in the wood and then slide them along. You will notice that your hand is carried upward toward the top of the tree. Another way of putting this is that when you place your right hand as shown, your thumb points upward. Had the grooves on the tree gone in the opposite direction, as seen on the right in the figure, then the thumb of your right hand would have pointed upward and trailed behind your hand as you slid your fingers along the grooves toward the top of the tree. You can test your comprehension of this distinction between right- and left- handedness by convincing yourself that if you put your *left* hand along the grooves on the tree shown in the right hand side of the figure, your thumb would be pointing upward.

For the sake of brevity, we will refer to the types of trees shown in figure 14–1 as "right-" and "left-handed" trees, respectively. Obviously, the distinction is a fundamental one in the sense that a twisted tree is either right- or left-handed. In another sense, however, the distinction is not fundamental. Both right- and left-handed trees are of the same species and could, in principle, grow to the same height and fill the same niche in the forest ecology. They might be thought of as being analogous to two houses that are identical in every respect except that one is painted blue and one red. If this is really true, then an important question arises. Why has nature bothered to create two

A deadfall of right-handed trees. Digging into the wood convinced us that the spiral goes all the way through. Custer National Forest, Montana.

versions of the same tree—versions that differ only in that one tree is the right- or left-handed version of the other? Although we were led to this question by the twisted tree alongside the trail, we shall see that right- and left-handedness constitute a deep and abiding puzzle in the universe.

But back to our trees. The first right-handed tree we saw didn't excite any response beyond a "Gee, look at that." As we continued to climb, however, we came across more and more of the twisted trees, and we soon noticed that all of them seemed to be right-handed. We then came to a large deadfall (see the photograph above). All of the trees in the fall were right-handed, and a few had decayed sufficiently for us to be able to dig into the material at the core. We found that the twisting was not a surface effect, but extended clear through the wood. This meant that the twisting is something that goes on throughout the life of the tree, not something that starts to occur when the tree reaches old age.

We began thinking about what could have caused the trees to twist. Because the twisting seemed to occur only near the timberline, we concluded that it must have something to do with

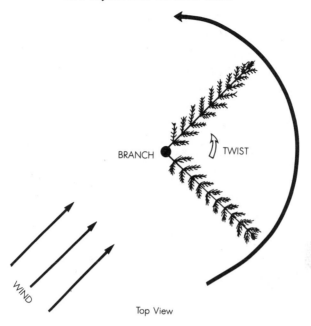

BRANCH ● TWIST

WIND

Top View

**FIGURE 14.2**

the stress placed on the tree by its environment. The existence of twisting was probably analogous to the deformed bones that we see in animals that have suffered severe vitamin deficiencies. But knowing that stress causes trees to twist doesn't tell us why they twist in a particular direction.

The first thing we considered was physical effects. For example, if a tree grew as shown in figure 14–2 above, in a region where the wind usually blew in one direction, and if there were more branches on one side of the tree than the other, you might expect the wind to exert a force that would twist the tree around. You often see trees sculpted into weird shapes along the seashore, where steady winds blow in from the ocean.

But this explanation simply didn't fit our data. For one thing, the twisting force of the wind on the tree in figure 14–2 would only be able to turn the tree partway around. If the tree were to be twisted a full 180°, for example, most of the branches would have been moved around to the other side of the tree and the force of the same wind would tend to unwind the tree. A second argument against wind as the cause of the twist is that the distribution of branches on a tree should be more or less random, so that there should be as many trees with more branches on one side as others with the branches on the opposite side. If the wind were the driving force, there should be as many left-handed trees as there are right-handed ones, a prediction that goes against the observed facts.

Well, what about genetics? Suppose there is something in the genes of these trees that predisposes them to twist to the right when they are stressed? This need not be an adaptive trait on which natural selection might act; it could be an accidental characteristic of the species. Even if right- and left-handed trees were equally distributed in the population in general, it could very well be that the first tree to colonize this particular valley was of the right-handed variety, and that what we were seeing was the result of the survival, some time in the past, of one particular right-handed seedling while another (perhaps left-handed) didn't manage to grow. That notion seemed to answer our question.

As we continued our climb, we congratulated ourselves on finding this neat explanation of the twisted trees. But soon we entered a grove of old trees and encountered the giant shown in the photograph opposite. Without a doubt, this tree was left-handed. During the rest of the day we saw others of this type, and eventually concluded that left-handed trees made up some 2 to 3 percent of our population.

The problem of the trees now became more difficult. We not only had to ask why nature chooses to introduce the distinction between right- and left-handed trees, but why almost all the trees are right-handed, and only a small minority left-handed. It would be easier to answer this question if every tree were of one type or the other; but what are we to do when there is a general rule which is broken every once in a while?

We didn't get any other ideas about the twisted trees during the hike, so when we got home I started asking around among members of the Forest Service. Everyone I talked to knew that such trees existed up near the timberline, but no one had a clue as to why they should be right- or left-handed. Still curious, I started asking my question whenever I met someone who studied the growth of trees or other plants. The answer was always the same. The twisting growth pattern occurs near the timberline, where the tree is near the limits of its range. The extra

A singular data point. Jeff Newmeyer stands next to our champion left-handed tree, growing in solitary splendor in a right-handed grove. Absarokee Wilderness Area, Montana.

stress placed on the organism presumably results in the abnormal growth; but other than a few comments about the role of the wind and one joking reference to the rotation of the earth, no one I talked to could offer any explanation at all about the direction of the twist.

Frankly, I was delighted by this turn of events. I happen to think the world is too full of things that can be explained. A little mystery here and there never hurt anything except, perhaps, the dignity of a scientist. Knowing that there are things in the world—humble though they may be—that aren't understood keeps us from getting too arrogant.

As I looked further into the question of the twisted trees, I discovered a number of other right- versus left-handed phenomena in nature. In England, for example, it is believed that the honeysuckle vine only grows in a right-handed spiral as it climbs a tree. A quick trip to my backyard in Virginia, however, convinced me that whatever it is that English honeysuckle does, its American counterpart will get up a tree any old way it can.

In the same vein, I have read that when the millions of bats that live inside Carlsbad Caverns in New Mexico come out of the cave every night, they sweep up into the sky following the path of a right-handed helix. I've noticed the same thing with hawks rising on thermal air currents (see chapter 1). They always seem to go around the thermal in a right-handed direction. Why either of these facts should be true is not known, although in both cases there are obvious advantages to having everyone moving in the same direction—it prevents traffic accidents. The question remains why the right-handed spiral is chosen over the left. Is some obscure law of nature involved, or is it just arbitrary and accidental, like the choice of driving on the right hand side of the road in most of the world?*

Perhaps the most striking example of handedness in nature lies not in the growth and behavior of living systems, but in the construction of the molecules without which life would be impossible. Most of us have heard of the "double helix," the shape of the DNA molecule that carries all genetic information. Since we are thinking about the problem of handedness, and since we know that these molecules have the shape of a helix, it is reasonable to ask whether the helices are right- or left-handed, and whether this handedness is the same in all living forms or not.

The answer is surprising. Not only DNA, but all molecules in living systems on earth, right down to the lowliest amino acid, are right-handed helices. This is truly surprising, because if you cook up a batch of amino acids in the laboratory, you will get roughly equal numbers of right- and left-handed molecules. When amino acids are found in exotic places like meteorites, the same applies—there are equal numbers of each

*When I was a student in England, I got into a typical student argument about the British convention of driving on the left. I pass along without comment, one reason I was given. "When a caveman entered a cave in which a fierce beast might be waiting," I was told, "he naturally turned so that his right, weapon-bearing hand pointed into the cave and his left side, with the vulnerable heart, was toward the wall. Since that time, all civilized people have driven on the left."

kind of molecule. It is only when you look into living systems that you find the preference for right-handed spirals.

Our present idea of the way life began on earth is that ultraviolet rays from the sun, acting on the methane, ammonia, and carbon dioxide that made up the earth's early atmosphere, created amino acids which fell and accumulated in the ocean. We are probably surer of this step on the road to life than we are of any other, simply because we can easily reproduce it in our laboratories. After a short time (geologically speaking) the oceans were a soup of amino acids, and these acids started the long process of chemical reactions that eventually led to more complex molecules.

The problem is this: we would expect the primordial soup to be made up of roughly equal amounts of right- and left-handed amino acids. The proteins and other molecules made from these basic building blocks, then, would also be expected to show no preponderance toward either right- or left-handedness. Why, then, are the living systems that evolved from this chemical stew composed of only right-handed molecules?

We do not know enough about the origins of life to give a definitive answer to this question, but there is one possible explanation that I find particularly appealing. Although we cannot recreate in our laboratories the steps that led from the first primitive reproducing system, we do know that before the earth was a billion years old, there were simple cells present. Imagine, for a moment, what the situation was like in the first tidal pool where such a cell appeared.

The pool would probably have been warm. Repeated advances of the ocean would have replenished its supply of amino acids, while evaporation would have served to boil the soup down to a rich broth. In the depths of the pool, sheltered from the ultraviolet rays of the sun, all sorts of chemical reactions took place. Eventually, through a process we still don't understand very well, a collection of organic molecules was able to segregate itself, take in simpler molecules from its surroundings, and reproduce its own particular pattern of atomic arrangements. The first cell was born.

Since the cell arose in a mixture that was made up of equal parts of right- and left-handed molecules, there was a 50 percent chance that it would be right-handed and a 50 percent chance that it would be left-handed. Suppose, for the sake of argument, that it was the former. The cell would feed on the broth in its pool, quickly turning most of the available nutrients into copies of itself. Some of these would be washed into the ocean, and some of those pioneers would find other concentrations of nutrients on which to feed. In a relatively short time, the world's oceans would be teeming with right-handed cells. As higher life forms evolved, capable of feeding on the primitive cells, they too would carry in their genetic material the memory of the right-handed molecules of their distant progenitor. If during this time a left-handed cell happened to evolve somewhere, it would find itself in a world where the resources were being rapidly consumed by its right-handed competitors. Furthermore, the right-handed cells would have been competing with each other for some time, and the laws of natural selection would have operated to turn them into much more efficient competitors than they had been at the start. The left-handed cell, as new kid on the block, wouldn't have a chance in this situation. There would be an almost insuperable advantage accruing to the cell that got there first.

In this scheme, then, the right-handedness of the molecules in earth's living organisms is just the result of an accident. We could equally well have turned out to be made of left-handed material. Our question thus goes back to the fortuitous collection of organic chemicals in that first tidal pool.

From these considerations, it is clear that handedness in nature is far from an unusual occurrence. The twisted trees that mystified me on that hike are just one example among thousands of natural preferences for right- or left-handed structures in nature. In most cases, the handedness seems to be an accidental quality having nothing to do with adaptation or survival.

Handedness as we have described it is only one instance of the sort of thing in nature that physicists refer to as "symmetry." One way to understand the concept of this symmetry is to imagine looking at something directly, and then looking at the same thing in a mirror. If a tree is growing straight up, with no twisting, then it makes no difference which way you see it— the tree appears the same. In the jargon of physics, we say that a straight tree is invariant under reflection in a mirror. If, on the other hand, we start with a right-handed tree, it will appear left-handed when viewed in a mirror. For, as everybody knows, when you look at your reflection, it is the left hand of the mirror person that moves when you move your own right. We then say that any image is reversed under reflection. Another word used to denote the behavior of things seen in mirror image is "parity." A straight tree, which looks straight in both the direct and the mirror image, is said to have positive parity, while twisted trees, where left and right interchange on reflection, are said to have negative parity.

The situations of greatest interest to physicists who study symmetry are those in which the parity operation does not affect the thing being viewed. Let's suppose, for example, that there is no difference between right- and left-handed trees in terms of the amount of lumber that one would expect to obtain. Suppose further that you are the owner of a logging company surveying slides of aerial photographs of a forest, trying to plan how to harvest the timber. Would it matter to you whether the slides had been put into the projector properly or had been reversed?

The answer to this question is obviously no. Since both right- and left-handed trees yield the same number of board feet of lumber, all you want to know is the total number of trees. And as the direct and reversed slides give the same answer for the total yield of the forest, you don't have to know which one you are looking at. In the language of physics, the number of board feet available is invariant under the parity operation.

There is another situation in which we might find the same invariance. Suppose that right- and left-handed trees were equally common, which is what you'd expect if handedness were a random property of each tree. In that case, you could look at a slide of the forest and count up the number of each kind of tree. You would find that there was no excess of right-handed trees over left-handed, or vice versa. This situation, too, is invariant under parity, as you can see if you imagine comparing the direct and reversed slides of the forest. You would note that what was a left-handed tree in one slide was right-handed in the other, but in each slide the two types would still be equal numbers. In this example, we say that parity is a good symmetry of the forest.

But in the real forest that we hiked through and described earlier, this statement is not true. Almost all the trees were right-handed, with only a smattering of the left-handed variety. The reversed slides would show a forest of left-handed trees with a smattering of right-handers, and we could easily distinguish between the direct and reversed situations. In the language of physics, we would say that in our real forest the symmetry is broken, or that the forest displays a broken symmetry.

The reason my physicist friend and I were surprised to find the broken symmetry was our expectation that handedness would be randomly distributed among the members of the tree population. We expected, in other words, that the forest would be invariant under the parity operation. The observation of many parts of nature shows that this expectation is wrong for living systems, and even for the molecules on which life is based. Given this state of affairs, we should not have been surprised by what we found. In fact, we ought to have been surprised had things been any other way.

From the physicist's point of view, the tantalizing question is why parity is broken so repeatedly in nature. We know that everything is composed of atoms, and we know that most simple atoms are like the straight trees in our example: they appear to be the same whether seen directly or in a mirror. More important, we know that the electrical force that governs the interaction of one atom with another makes no distinction between right and left; it is, in the language we have introduced, invariant under parity. How, then, can invariant forces acting on invariant molecules produce something like DNA, or the twisted trees, which are most emphatically not invariant?

While you are mulling over this point, let me mention another broken symmetry in nature—the symmetry between matter and antimatter. Beginning with an experiment done in 1932, physicists have discovered that for every particle that exists in normal matter, it is possible to create another type of particle with the same mass as the original but with everything else reversed. For example, the normal electron has a negative electrical charge. The reversed particle, called the positron, has the same mass as the electron but a positive electrical charge. The positron (which was discovered in the experiment mentioned above) is one example of an antiparticle. If a particle and an antiparticle come together, they undergo a process called annihilation, in which the two have their energy converted into either radiation or other types of particles.

As was the case with parity, the laws of nature apparently are such that at the most fundamental level, they are invariant under the interchange of particles with antiparticles. Such an interchange, called "charge conjugation" in the language of physics, is the analogue of the interchange of right and left when an image is reversed. The obvious question: If nature seems to be invariant under this sort of interchange, why is the world made completely of matter, with only the smallest addition of antimatter? This question is, of course, the exact analogue of our question about how parity invariant atoms and forces can produce the right-handed forest we saw.

The so-called antimatter problem has occupied the attention of physicists and astronomers for decades. We know that the earth is made entirely of matter—any pieces of antimatter that might have been around would long ago have undergone annihilation and disappeared. When we examine meteorites that come to us from other parts of the solar system, we find that they too are composed entirely of matter. We are entitled to conclude that our own solar system contains a predominance of matter.

We can say the same about our own and nearby galaxies. Cosmic rays are emitted by stars throughout space, and they rain down continually on the earth. Most of these particles come from stars in our own galaxies, but a small percentage originate in other galaxies. When we look at the cosmic rays, we find them to be made up almost entirely of matter, and the small admixtures of antimatter particles we do see are made in collisions between cosmic rays in the depths of space.

For a while, astronomers thought that the universe might have equal proportions of matter and antimatter, but that the antimatter had somehow become segregated from the matter. If this were true, then annihilation would occur only on the boundaries between regions of matter and antimatter. In such a case, the annihilation regions would produce copious X-rays, which we would have been able to see as soon as we had X-ray detectors above the atmosphere (see chapter 10). But none were seen, and the only conclusion we can draw from this fact is that despite the apparent symmetry of the laws of nature under interchange of particle and antiparticle, the actual universe exhibits a high level of symmetry breaking.

The disappearance of the antigalaxies from the scientific scene was something of a loss, at least in a literary sense. How many great science fiction epics were based on conflicts between us and some sort of anti-world? My own favorite literary excursion in this realm was a poem that appeared in *The New Yorker* many years ago, beginning:

> *Well up beyond the tropostrata*
> *There is a region stark and stellar*
> *Where, on a streak of anti-matter,*
> *Lived Dr. Edward Anti-Teller.*

In a more serious vein, the study of the breaking of symmetries has been one of the most fruitful and dynamic fields of physics during the last decade. One aspect of symmetry breaking, typified by the twisted trees and the parity operation, has been known for quite some time, although its full significance only recently has been widely appreciated. Another aspect, typified by the antimatter problem, was understood only at the end of the 1970s. The key point in understanding both types of problems is that human beings have an almost infinite capacity for believing that the universe is actually the way they think it ought to be.

Let me give you one example. Earlier, I raised the question of how DNA could have a definite handedness when its constituent atoms and the forces holding it together did not. The unspoken assumption there was that if the constituents and forces were invariant under parity, the final entity assembled by those forces from those constituents ought to have the same property. But what is the basis for believing that assumption to be true? For me, it was just a "gut feeling," nothing more.

There is at least one familiar example to show that this assumption, this expectation of how things ought to be, must be wrong. Think of an ordinary roulette wheel that's been painted white all over so that the numbers are no longer visible. The wheel will obviously look the same in a mirror as it does under direct sighting. Similarly, the small metal sphere that rolls inside the wheel is invariant under parity. When the wheel is spun and the ball thrown, all the forces present (gravity and the electrical forces between atoms) are also invariant under parity. Yet if we wait for a while, the ball will fall into one of the slots on the wheel, and the wheel itself will come to a halt.

As soon as the ball goes off center, the system is no longer invariant under parity. If the ball is on the right-hand side of the wheel when we look at it directly, it will be on the left when we look in the mirror.

From this example, we learn that our expectation that invariant constituents and invariant forces always lead to invariant final results is altogether naive. Nature just doesn't behave that way, much as we may think it ought to. It offers many examples of this sort of symmetry breaking, and even if all the forces in nature were invariant under parity (which they're not), DNA would still be a right-handed helix.

As for the breaking of symmetry between matter and anti-matter, its explanation is not nearly so simple and obvious as our account of handedness. The idea that nature should be the same when particles and antiparticles are interchanged is not an obvious one—no one has gut feelings here. It just so happened that in the early days of the study of elementary particles, all the reactions that were seen had this invariant property. The idea that the symmetry had to be universal simply seeped into the accepted wisdom of physics, without being subjected to a rigorous examination.*

In 1964 two Princeton physicists, James Cronin and Val Fitch, announced the results of the first of a series of experiments which would eventually show that the symmetry between matter and antimatter that everyone expected to be universal in nature was broken. It wasn't broken by much, of course, or we would never have expected the symmetry to hold in the first place. What they found was one system (called the K mesons) where there were slightly more positrons than electrons produced when the particles decayed.

I think of this discovery of the low level of symmetry breaking by Fitch and Cronin (for which they received the Nobel Prize in 1980) as being analogous to our discovery of the occasional left-handed tree in the mountains. Both discoveries show that

*Much the same thing happened with regard to the idea that the fundamental forces were invariant under parity. For the story of the "downfall of parity," you might want to consult Martin Gardner's *The Ambidextrous Universe* (Scribners, 1979).

nature is not as symmetrical as we would expect at first glance.

On the assumption that the universe has always been as we see it today, the tiny effect discovered by Fitch and Cronin cannot possibly explain the observed predominance of matter over antimatter. At the start of the Big Bang, however, when temperatures were much higher than they are now, the universe was much more sensitive to small symmetry-breaking effects than it is now. The situation is similar to the development of a human life: a cell division just after fertilization can have a crucial effect on the development of the body, while a single cell division in an adult is unlikely to matter very much.

What we believe happened in the early universe was this: some $10^{-35}$ seconds after the expansion associated with the Big Bang, there were as many particles as antiparticles. At that time, the sort of small asymmetry observed by Fitch and Cronin played a crucial role. Although there were equal amounts of matter and antimatter present at first, particles starting to decay produced a situation with slightly more matter than antimatter. Perhaps there were a billion and one protons for every billion antiprotons. As time went on, the protons and antiprotons collided and annihilated until, in our present era, all that is left is the odd proton created in the decays during the first second after the Big Bang.

It seems, then, that when we observe nature, we have to be aware that symmetries exist; but we must also realize that sometimes the most interesting lessons are to be learned from seeing what happens when the symmetry is broken. Whether these symmetry-breaking effects occur on the macroscopic level, as with the trees, or at the level of elementary particles, as with antimatter, they have already played an important role in shaping the universe we live in.

# Index